MW00522985

The Greatest Minds in Science

METRO BOOKS
New York

An Imprint of Sterling Publishing Co., Inc.
1166 Avenue of the Americas
New York, NY 10036

ISBN 978-1-4351-6305-8

For information about custom editions, special sales, and premium and corporate purchases, please contact Sterling Special Sales at 800-805-5489 or specialsales@sterlingpublishing.com.

Manufactured in China

2 4 6 8 10 9 7 5 3 1

www.sterlingpublishing.com

Credits: Illustrations by Eva Tatcheva
Design by Elwin Street Productions Limited

The Greatest Minds in Science

50 Ideas That Changed the World

Peter Moore and Mark Frary

METRO BOOKS
New York

Contents

Introduction

As a way of thinking, science has produced remarkable results. Its method of attempting to make sense of the world in which we live by not only asking questions but going on to perform carefully organized experiments has let humans explore inside minute atoms as well as probe the vastness of space.

It wasn't always this way. Ancient thinkers used to try to work out how the world and the heavens worked by asking questions and then using processes of dialogue and argument to come to answers. Few thought that studying physical objects would provide anything useful and until the seventeenth century, fewer still made any attempts at accurate measurements. There are, however, always exceptions, and a few key individuals led the way in making some staggeringly detailed studies of the most vital aspects of our universe.

This book highlights fifty of the key scientists who have made these remarkable discoveries, in each case focusing on the major contributions that they have made. We learn how sequences of scientists and their discoveries have led to ever fuller understanding of concepts such as the make-up of gases, the development and implementation of electricity as a power source, and the fight against the microscopic bugs that cause infection. From there you will see how these theories have had long-lasting influence, affecting our lives to this day.

Progress in science has been huge, and the spin-offs immense, but this book makes the key discoveries accessible and fascinating, as well as reminding us that, although new ideas and insights are hard to predict, they may be just around the corner.

150 CE

1525

1550

1575

1600

1625

1700

1800

1900

1925

1930

1935

1940

1980

Ptolemy *Geographia* (ca. 150 CE)

Nicolaus Copernicus *De revolutionibus orbium coelestium* (On the Revolutions of the Celestial Spheres) (1543)

Tycho Brahe *De nova stella* (On the New Star) (1573)

Johannes Kepler *Mysterium Cosmographicum* (The Cosmographic Mystery) (1596)

Galileo Galilei *Sidereus Nuncius* (Sidereal Messenger) (1610)

Fredrich Bessel *Fundamenta Astronomiæ* (1818)

Abbé Lemaître "Un univers homogène de masse constante et de rayon croissant rendant compte de la vitesse radiale des nébuleuses extragalactiques" (A homogeneous universe of constant mass and growing radius accounting for the radial velocity of extragalactic nebulae) (1927)

Karl Jansky "Radio waves from outside the solar system" (1933)

Edwin Hubble *The Observational Approach to Cosmology* (1937)

Stephen Hawking *A Brief History of Time* (1988)

Astronomy

The night sky, that canvas of black velvet jeweled with twinkling diamonds, has long inspired deep thinkers. From the earliest observers of the wandering stars of our skies to modern-day astrophysicist scientists have delved into the field of astronomy in a bid to understand Earth's place in the universe.

The Solar System

Earth is a small but remarkable planet set in the middle of a small and probably unremarkable solar system, tucked into the corner of a medium-size galaxy, which is just one of billions in the universe. Working this out was no mean feat.

When the famous astronomer Nicolaus Copernicus (1474–1543) viewed the heavens he came to a radical conclusion: rather than accepting the ancient view that everything revolved around the Earth, he showed that another explanation for our observations was that the sun was the central object. When Johannes Kepler produced final proof for Copernicus's theory in 1621, the concept of a "solar system" became irrefutable.

Even so there was much to discover. At the beginning of the seventeenth century, astronomers had only been able to spot eight bodies that moved across the skies. These were the sun, Mercury, Venus, Earth and its moon, Mars, Jupiter, and Saturn. Uranus wasn't known of until William Herschel spotted it in 1781, Neptune was first seen by Johann Gotfried Galle in 1846, and Pluto by Clyde Tombaugh in 1930.

During this period various observers were beginning to detect "moons" orbiting many of the planets. In 1610, Galileo spotted Callisto, Europa, Ganymede, and Io all orbiting Jupiter. It was a remarkable feat of observation, even though he managed to miss the other twenty-one bodies that orbit that faraway planet.
Indeed, each time we get a better view of the solar system we seem to find more objects to name and study. The twin Voyager satellites that were launched in 1977 have gradually made their way through the solar system, making close encounters with many of the planets as they travel. Between 1985 and 1989 Voyager 2 beamed back information about sixteen newly identified major bodies in the solar system, bringing the total known so far to seventy-one.

"The sun, with all those planets revolving around it and dependent on it, can still ripen a bunch of grapes as if it had nothing else in the universe to do."

–Galileo Galilei

At a distance of over three and a half billion miles, Pluto is the farthest planet from the sun. In fact, given its small size and highly elliptical orbit, some people question whether it should really be classified as a planet at all, even if only as a drawf-planet, but given that it orbits the sun it is undoubtedly part of the solar system.

As well as beginning to realize the intricacies involved in our own solar system, astronomers have started to see that this is just the start. Each star in the sky represents another sun, many of which we now know are orbited by planets of their own. Our solar system rests in one of the spiraling arms of a medium-size galaxy that we call the Milky Way. It is about 100,000 light years across and contains approximately100 million stars. The sun orbits around the center of the galaxy, making one rotation every 225 million years. It turns out that this galaxy is just one of billions of others found within the universe, some of which contain up to three trillion stars. On this scale, our solar system is tiny, for all that it is so important for us.

Claudius Ptolemaeus ("Ptolemy")

Mapping the Known World

Born
Ca. 90 CE
Alexandria, Egypt

Died
Ca.168 CE
Alexandria, Egypt

At the time when Ptolemy was working, the pervasive view was that Earth was at the center of the universe. For some, this was based on religious belief, for others the conclusion was drawn from philosophical arguments. Ptolemy therefore set out to pull together all previous observations, calculations, and theories, and show how such a geocentric view might work.

When astronomers looked at planets they became aware that they didn't just move smoothly through the sky night after night, but that there were times when they appeared to stand still for a moment, or even move backward. Ptolemy's explanation was that while the planets orbited Earth in circles that had centers close to Earth's center, they also moved in what he called epicircles.

To do his work, Ptolemy developed mathematical ways of relating lines and angles, and one of his findings has been handed down as Ptolemy's theorem. This shows that there is a fixed set of relationships in lengths of lines of a four-sided box drawn within a circle. The theory now underpins much current trigonometry.

The system drew heavily on the work of Greek and Babylonian astronomers such as Hipparchus (ca. 190 BCE–ca. 120 BCE) and required eighty different orbits; the combination of these orbits produced the pattern of behavior seen from Earth. This model of the universe remained unchallenged until the Polish scholar Nicolaus Copernicus (1473–1543) published his heliocentric view in 1543.

In addition to his work in astronomy, Ptolemy was a pioneering map-maker. Convinced that Earth was a sphere, he developed geometric methods of projecting a sphere onto a flat surface. He also included coordinates of latitude and longitude for every feature drawn on the map, allowing anyone to reproduce it at any scale they wished. By the year 1500 his book *Geographia* was so far ahead of anything else that it remained the principal work on the subject. The book was partly responsible for sending Christopher Columbus (1451–1506) sailing west in search of the Indies. Columbus thought that the journey would be short, because on Ptolemy's maps there was only a short span of ocean, but Ptolemy had underestimated the size of the Earth and had overestimated the size of Asia.

The groundbreaking maps created by **Ptolemy**, complete with latitude and longitude coordinates, cover about a quarter of the world's surface, from the Canary Islands in the west to China in the east, and from the Arctic to equatorial Africa.

Nicolaus Copernicus
Shifting Our View of the Earth

The Greek astronomer Aristarchus, in the third century BCE, was already proposing the theory of heliocentrism—the idea that the sun is at the center of the solar system. However, it was not until Catholic cleric Nicolaus Copernicus arrived on the scene with a mathematical model for how such a solar system might work that the idea gained acceptance.

At the time of Copernicus, there were two existing theories for how the universe operated. The first, by Aristotle (384–322 BCE), was based on the idea that the stars were located on a fixed sphere, or a series of fixed spheres, with Earth at their center; the second, due to Ptolemy (100–170 CE), was also based on spheres but added in complicated epicycles to explain the observed motion of some planets, which appeared to move backward in their orbits at times.

In 1497, Copernicus was made a canon at the cathedral of Frombork after several years at a number of universities across Europe studying the arts, medicine, and canon law; Copernicus's fascination with astronomy had started at the University of Krakow. In the same year that he joined the clergy, he made observations of the moon passing in front of the bright star Aldebaran in the constellation of Taurus, and his observations made him begin to doubt the idea of an Earth-centered universe.

In the following years, Copernicus made a series of meticulous observations of the sun and planets, which supported his idea that geocentric views of the universe were wrong. He published a short manuscript, shared only among his close acquaintances, known as the "Commentariolus," extolling these views. He soon began work on a more detailed model of heliocentrism but it would be more than thirty years before it saw the light of day.

It is often believed that the Catholic Church was against the heretical idea of a sun-centered universe but this is far from true. Pope Clement VII heard about Copernicus's work in the 1530s and wrote to the astronomer to encourage him to publish but to no avail. In the end it was the persuasive arguments of Copernicus's pupil Rheticus that convinced him to finally publish, in 1543, *De revolutionibus orbium coelestium*, the masterpiece that set out a mathematical model of a universe in which Earth is relegated to a lesser role.

Born
1473, Torun, Poland

Died
1543, Frombork, Poland

Although it was not **Copernicus** who initially proposed the idea of the sun being at the center of the universe, it was he who proved that it made more sense than existing Earth-centered models.

Tycho Brahe
Breaking the Celestial Sphere

"There is one heaven . . . ungenerated and eternal." These are words from the astronomical writings of Aristotle and outline the view that was to remain unchallenged for the next eighteen centuries. Until, that is, Danish astronomer Tycho Brahe, using the most advanced astronomical instruments of the pre-telescope era, built to his own designs, observed an event that was to change everything.

The constellation of Cassiopeia is one of the most recognizable in the night sky, thanks to its distinct W of bright stars. In the year 1572, this W was suddenly joined by something unexpected. Writing in the book *De Nova Stella*, Brahe explained his shock: "On the 11th day of November in the evening after sunset, I was contemplating the stars in a clear sky. I noticed that a new and unusual star, surpassing the other stars in brilliancy, was shining almost directly above my head; and since I had, from boyhood, known all the stars of the heavens perfectly, it was quite evident to me that there had never been any star in that place of the sky." What Brahe had seen was a supernova, the last gasp of a dying star when its stock of nuclear fuel runs dry. It challenged the prevailing view of the perfection of the heavens.

Brahe had been drawn to astronomy by a 1560 solar eclipse and continued to study it in his spare time at universities in Germany and Switzerland. On his return to Denmark, he became absorbed in building highly accurate astronomical measuring apparatus such as quadrants, sextants, and armillary spheres.

Yet the 1572 apparition changed everything. The publication of *De Nova Stella* was a sensation across Europe, showing that the long held Aristotelian view of the unchanging heavens was untrue. His services as an astronomer were suddenly in high demand and the Danish King offered him the island of Hven on which to build the world's greatest observatory at a cost amounting to five percent of the country's gross national product.

With the best instruments money could create, he made meticulous observations of the stars, planets, and comets. His observations of the so-called Great Comet of 1577 showed that comets were not atmospheric phenomena as most people believed but were far distant from Earth. These observations, and those he made of the planet Mars, were eventually used to show that heavenly bodies moved in elliptical, not circular orbits, shattering forever the illusion of celestial perfection and laying the foundations for Newton's theories of gravity.

Born
1546, Scania, Denmark [now part of Sweden]

Died
1601, Prague, Czech Republic

Brahe made observations of the stars and planets that were a hundred times more accurate than those of his predecessors, yet the telescope had not yet even been invented. His observations of the planet Mars, in particular, led directly to Newton's theories on gravity.

Galileo Galilei
Observing the Universe

When Galileo Galilei headed off to Padua to study, medical education was still firmly based in the teachings and writings of ancients such as Galen and Aristotle, with little room given for injecting any new ideas. It is hardly surprising then that Galileo soon switched his attentions to the less restrictive world of mathematics.

Galileo turned his mind to studying gravity. To slow down the action of gravity in order to better observe it, he ran a ball down a slope, rather than simply dropping it. He found that the time for the ball to travel along the first quarter of the track was the same as that required to complete the remaining three quarters. He therefore realized that the ball was constantly accelerating. By repeating the experiment hundreds of times to see whether the same thing happened on each occasion, he founded an important aspect of scientific method.

The seed of another of his discoveries started when he was distracted by a lantern swinging at the end of a chain one day in Pisa cathedral. He went on to realize that a pendulum takes the same amount of time to swing from side to side whether it is given a small or a large push. The frequency only changes if you alter the weight or change the length of the supporting rope.

Galileo also developed a two-lens telescope and started studying the moon, sun, and planets. In so doing, he came across wonders that he described as "never seen from the beginning of the world" such as sunspots and planet satellites. He then began to look into the work of the Polish-born medic, lawyer, and astronomer Nicolaus Copernicus (1473–1543), who in his 1543 landmark publication had said that the planets, including Earth, rotated around the sun; Earth was not the center of the universe.

Intrigued by this possibility, Galileo looked for evidence. He noticed that Venus went through phases, much like those of the moon. From this he deduced that Venus must be orbiting the sun. While he was right in his prediction he didn't have conclusive proof, and because it contradicted mainstream scriptural teaching, the Catholic Church instructed him to stop talking about it. Instead he wrote a book presenting his ideas entitled *Dialogue Concerning the Two Chief World Systems*, which included a thinly veiled insult toward the pope, Urban VIII. The result was Galileo being put under house arrest for the rest of his life.

Born
1564, Pisa, Italy

Died
1642, Arcetri, Italy

One of the first to use telescopic investigation, **Galileo** was the first person to see sunspots, the first to identify the four main satellites of Jupiter, and the first to record that the surface of the moon had mountains and craters.

Johannes Kepler
Understanding Planetary Motion

Chiefly remembered for discovering three key laws of planetary motion and making some incredibly accurate astronomic tables, Johannes Kepler also made important discoveries in the fields of optics and mathematics.

Intending to be ordained as a priest, Kepler went on to see his scientific research as a way of fulfilling a Christian duty to understand the works of God, saying at one point that he was merely thinking God's thoughts after him. He was convinced that God had created the universe according to some mathematical plan, and that mathematics would be the route to understanding it.

At Tübingen, Kepler started studying under astronomer Michael Maestlin (1550–1631). To teach some more advanced astronomy Maestlin introduced Nicolaus Copernicus's (1473–1543) recently published idea that the sun, not the Earth, was the center of the universe—a thoery to which Kepler was instantly drawn. He first developed a complex argument, suggesting that the paths of the planets could be predicted by calculating the sizes of a series of spheres, cubes, and tetrahedrons nested inside each other. The arguments seem curious to twenty-first-century thought, but the results closely correlated with astronomical measurements.

Looking more closely, however, Kepler discovered that Mars's orbit was elliptical rather than circular, with the sun one of the foci of the ellipse. He produced over a thousand sheets of mathematical workings before reaching this conclusion, and then proceeded to study the other planets. The fact that all the planets' orbits turned out to be ellipses became known as Kepler's first law of planetary motion. His second law was the realization that a line joining a planet to the sun swept out equal areas in equal times as the planet traveled around its orbit.

While Copernicus had presented the initial idea, and Galileo Galilei (1564–1642) had added some more observational data, it wasn't until Kepler had finished his work that for the first time there was mathematical and scientific proof that the planets orbited the sun. The universe hadn't changed, but our understanding of it had.

Born
1571, Weil der Stadt, Württemberg [now Germany]

Died
1630, Regensburg [now Germany]

As a result of his astronomical observations, **Kepler** also became interested in optics, discovering how to make a telescope that used two convex lenses. His design became so widely used that it is simply called an astronomical telescope.

Fredrich Bessel
Charting the Heavens

In 1804, at the age of twenty, Fredrich Bessel wrote a paper on Halley's Comet in which he calculated its orbit using observations made by Thomas Harriot (1560–1621) in 1607. He sent his paper to leading comet expert Heinrich Olbers (1758–1840) who subsequently asked Bessel to make further observations and also asked him to consider becoming a professional astronomer.

Bessel's response was to study celestial mechanics, firstly at the privately owned Lilienthal Observatory near Bremen and then at the newly built Observatory at Königsberg, where he remained for the rest of his life.

It was in Königsberg that Bessel determined the positions and relative motions of more than 50,000 stars. His starting point was the data of English Astronomer Royal James Bradley (1693–1762). This work produced a system of predicting the relative positions of stars and planets. Bessel was one of the first astronomers to realize the importance of working out how many errors were involved in taking measurements. By working out all the sources of error that could be generated by Bradley's instruments, he created a much more accurate set of results. This enabled him to state the positions of stars on particular dates and eliminate from his reckonings such factors as the effects of Earth's motion.

By eliminating all sources of error—meteorological, optical, and mechanical—Bessel was able to obtain results of astonishing accuracy from which a great deal of new data could be extracted.

In 1830 Bessel published data showing the positions of thirty-eight stars over the hundred-year period from 1750 to 1850. He spotted that two stars, Sirius and Procyon, moved somewhat erratically, and deduced that this variation in their movement must be caused by the tug of previously unseen companion stars orbiting them. He announced in 1841 that Sirius had an unseen ("dark") companion star. It was not until ten years later that the orbit of the companion star was computed, and astronomers finally managed to see it in 1862, sixteen years after Bessel's death.

Bessel also devised a method of mathematical analysis involving what is now known as the Bessel function. Helping to analyze the way that the gravitational forces of three objects interact with each other as they move, the function has become indispensable in applied mathematics, physics, and engineering.

Born
1784, Minden, Brandenburg [now in Germany]

Died
1846, Königsberg, Prussia [now Kaliningrad, Russia]

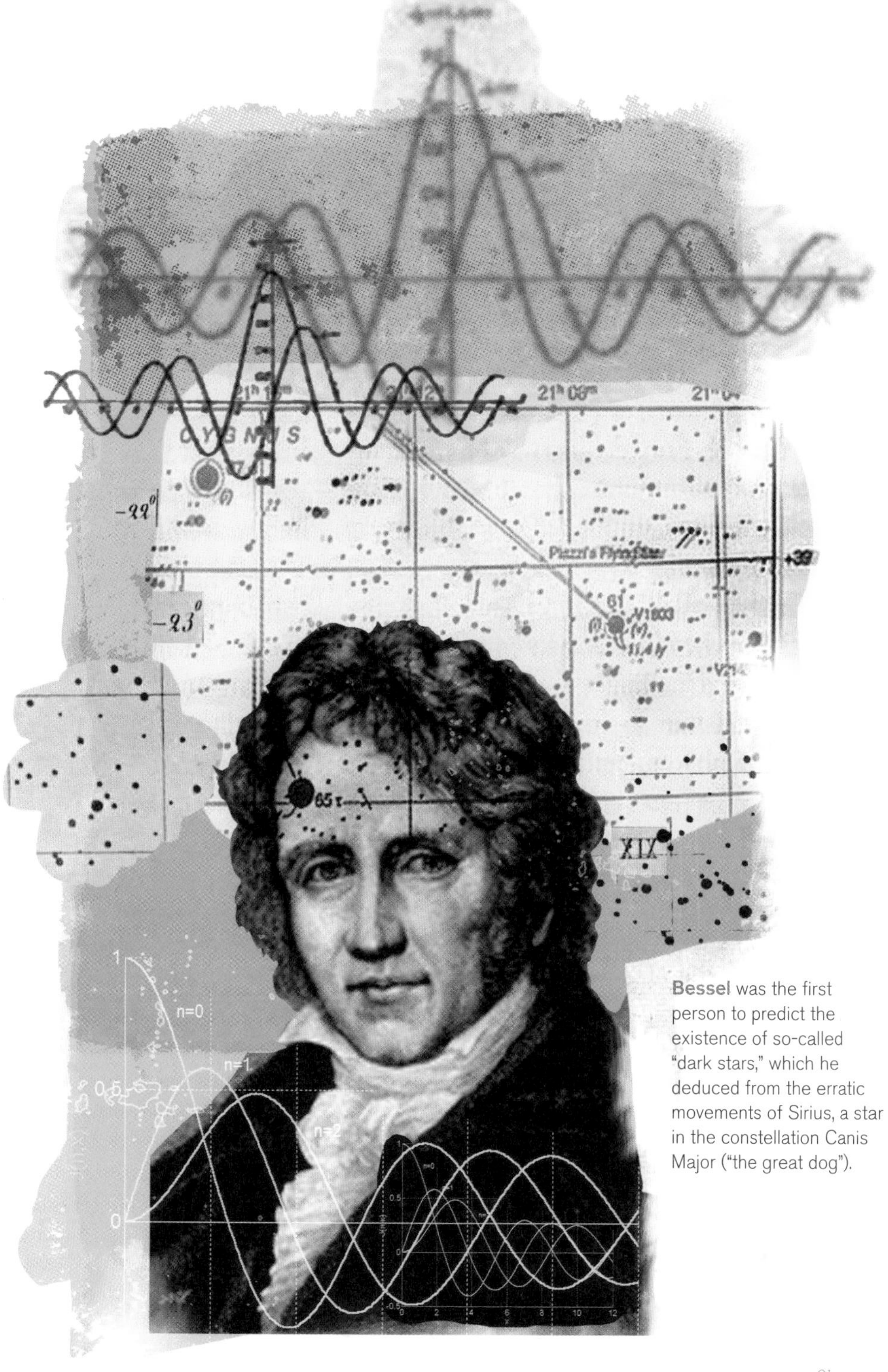

Bessel was the first person to predict the existence of so-called "dark stars," which he deduced from the erratic movements of Sirius, a star in the constellation Canis Major ("the great dog").

Edwin Hubble

Discovering the Expansion of the Universe

Born
1889, Marshfield, Missouri

Died
1953, San Marino, California

In the 1920s most of Edwin Hubble's colleagues believed the Milky Way galaxy made up the entire cosmos. But peering deep into space, Hubble realized that the Milky Way is just one of millions of galaxies, and that these galaxies are all rushing away from each other.

Having studied science in Chicago and Oxford, Edwin Hubble started to examine the stars at Yerkes Observatory in Wisconsin before moving on to the prestigious Mount Wilson Observatory in California, which housed the world's most powerful telescope. The main focus of his attention was on strange, fussy clouds of light called "nebulae."

At Mount Wilson, Hubble found himself working alongside Harlow Shapley (1885–1972), an astronomer who had recently measured the size of the Milky Way. Using bright stars called Cepheid variables as standardized light sources, he had calculated that the galaxy was 300,000 light-years across—ten times bigger than anyone had thought. Shapley was convinced that the Milky Way contained all of the stars and matter in the universe—that there was nothing beyond it. Shapley believed that the luminous nebulae that interested Hubble were just clouds of glowing gas, and they were relatively nearby.

In 1924 however, Hubble spotted a Cepheid variable star in the Andromeda nebula, and using Shapley's technique showed that the nebula was nearly a million light-years away—a fact that placed it way outside the Milky Way. We now know that this is the nearest of tens of billions of galaxies.

This alone didn't satisfy Hubble's curiosity. As he studied Andromeda, he realized that the light coming from it was slightly redder than he would have anticipated. The effect is similar to listening to the siren of a moving police car. As it approaches, the tone goes higher, and as it passes the tone drops. A shift toward red is equivalent to a drop in tone. The most likely cause of this so-called red shifting was that the galaxies were moving away from the Milky Way—from our own galaxy. By measuring the shift in all the nebulae he could find, Hubble came to realize that the farther away a galaxy is from Earth, the greater the red shift—in other words, the faster it is moving away from us. The explanation was extraordinary: The entire universe is expanding.

The observations that **Hubble** made about the expanding universe—sometimes referred to as "Hubble's law"—were ground-breaking and went some way to explaining the Big Bang theory.

(Abbé) Georges Lemaître

Understanding the Origins of the Universe

Born
1894, Charleroi, Belgium

Died
1966, Louvain, Belgium

Originally a priest, Georges Lemaître's interest in astronomy stemmed from his studies about Creation, which, combined with his scientific and mathematical work, led him to propose the Big Bang theory.

When Lemaître started studying the universe, most scientists thought that it was infinite in age and basically unchanged in its appearance—that it had always been there. Lemaître wasn't convinced. He reviewed Einstein's general theory of relativity and agreed that the universe had to be either shrinking or expanding. However, while Einstein had added a cosmological constant to let his equations work in a stable universe, Lemaître decided that the universe was expanding. This theory fit with early observations of a red shift in color of light from far-off galaxies, which he thought could be explained if these galaxies were moving away from us. Lemaître published his calculations and reasoning in 1927, but few people took any notice.

Two years later Edwin Hubble confirmed the existence of the red shift. Lemaître sent a copy of his ideas to Arthur Eddington (1882–1944), a member of the Royal Astronomical Society in London, who realized that Lemaître had bridged the gap between observation and theory, and the Royal Astronomical Society subsequently published an English translation of Lemaître's paper in 1931. Even so, most scientists found the idea of the universe having a beginning repugnant, and Cambridge astronomer Fred Hoyle (1915–2001) coined the term "Big Bang" as a joke.

Lemaître knew that there were still problems with his theory. For example, he predicted that the universe had expanded at a steady state, but this meant that it was expanding too quickly for the stars and planets to form. Lemaître solved this by using a variation of Einstein's cosmological constant to speed up the expansion of the universe over time.

Confirmation of the Big Bang took another thirty-three years. In 1964, Arno Penzias (1933–) and Robert Wilson (1936–), working at Bell Laboratories in New Jersey, were studying the interference to radio signals captured from echo balloon satellites by a sensitive six-meter-wide antenna. The interference did not vary over the course of a day and appeared to come from all directions of the sky, meaning that its origin was likely to be beyond the galaxy. Penzias and Wilson, working with a team at nearby Princeton, identified the source as the cooled microwave remnants of the Big Bang. Lemaître's theory was finally secure.

Lemaître believed that his Big Bang theory helped to explain that the red shift in color of light from far-off galaxies happened because the galaxies were moving away from us.

Karl Jansky
Pioneering the Study of Radio Astronomy

Born
1905, Norman, Oklahoma

Died
1950, Red Bank, New Jersey

Scientific advances are often made from unexpected discoveries; penicillin, X-rays, and the smallpox vaccine are among just some of the "accidents" that have led to monumental progress. Another such "accidental" discovery of huge importance was made by radio engineer Karl Jansky when looking into possible sources of short wave radio interference.

Having graduated with a degree in physics from the University of Wisconsin, Jansky accepted a job as a radio engineer at the Holmdel site of telecommunications company Bell Labs. This was still in the early years of radio; Marconi had sent his first signal just over thirty years earlier.

Bell asked Jansky to investigate possible sources of interference to "short" waves (those with wavelengths of tens of meters), which could be used for transatlantic radio telephone services. Jansky built a bulky antenna—which became known as Jansky's merry-go-round because it could be rotated on a turntable—and he set about identifying sources of interference.

Jansky was quick to pinpoint two sources of static hiss: local thunderstorms and, slightly harder to work out, thunderstorms that were beyond visual range. However, another source of static, a low hiss, defied Jansky for a year. What was curious about the signal was that it seemed to vary in intensity with peaks occurring once a day. As a result of this daily cycle, Jansky started to believe that the sun may be the source of the interference. However, after months of observations, he noticed that instead of a cycle of 24 hours, the source actually had a cycle of 23 hours 56 minutes, suggesting an origin beyond the solar system.

Jansky's big leap was in recognizing that the interfering radio waves came from the heart of our galaxy, the Milky Way, in the direction of the constellation of Sagittarius. His findings were published in a paper called *Electrical Disturbances Apparently of Extraterrestrial Origin*, the word "Apparently" included as a sop to his Bell superiors who were worried that they might be ridiculed. They need not have worried. Although Jansky was unable to continue his research into the Milky Way because his superiors wanted him to concentrate on telecommunications problems, he laid the foundations for the important subfield of astronomy that became known as radio astronomy.

Jansky did not continue his studies of radio astronomy after his initial discoveries, but a young radio operator from Illinois, Grote Reber (1911–2002), having read Jansky's paper, went on to build the world's first radio astronomy dish in 1937.

Stephen Hawking
Understanding the Secrets of the Universe

Many people question whether humans will ever understand something as vast as the universe. Best-selling author and astrophysicist Stephen Hawking has gone further than most in developing theories that make sense of current data.

Hawking has spent most of his academic life at the University of Cambridge, focusing largely on developing Albert Einstein's theory of general relativity, which had introduced the concept of space-time. In special relativity and general relativity, time and three-dimensional space are treated together as a single four-dimensional concept called space-time. A point in space-time is called an event, and an event must have four reference points: three positional coordinates and time.

Between 1965 and 1970, and working in collaboration with Roger Penrose (1931–), Hawking devised new mathematical techniques to study space-time. He then went on to apply this to the study of black holes, which appear to have been formed by stars collapsing in on themselves and becoming so dense, and their gravitational fields so strong, that, like virtually everything else, light cannot escape their pull. By combining quantum theory—the physics of the very small—and general relativity, Hawking indicated that black holes can emit radiation. From then on he started working to try and roll quantum theory and general relativity into what was hoped would become a "grand unified theory."

A sign that this may be possible came when Hawking investigated predictions about the creation of the universe that flow from these two theories. He started by calculating that following the Big Bang many objects the size of a proton would have been created. While a proton is incredibly small, these particles might have a mass of as much as ten billion tons. The large mass of these mini black holes would give them huge gravitational attraction and they would therefore be governed by general relativity, but their small size would make them also governed by laws of quantum mechanics.

Mini black holes are yet to be observed but Hawking's thoughts on space-time have moved us one step further down the road to a grand unified theory of the universe. Yet this theory is unlikely to be anything philosophical but rather a series of mathematical equations. In Hawking's own words, "What is it that breathes fire into the equations and makes a universe for them to describe?"

Born
1942, Oxford, England

Hawking combined quantum theory and general relativity to show that black holes emit an energy—now called "Hawking radiation"—rather than simply suck things in, as was previously believed.

1600

Isaac Newton *Philosophiæ Naturalis Principia Mathematica* (Mathematical Principles of Natural Philosophy) (1687)

1700

Alessandro Volta *De vi Attractiva Ignis Electrici* (On the Attractive Force of Electric Fire) (1769)

1800

Michael Faraday *Experimental Researches in Electricity* (1839–1844)

1850

James Clerk Maxwell *A Dynamical Theory of the Electromagnetic Field* (1865)

1875

1890

Heinrich Hertz *Electric Waves: being researches on the propagation of electric action with finite velocity through space* (1893)

Wilhelm Röntgen *Über eine neue Art von Strahlen* (On A New Kind of Rays) (1895)

1895

Max Planck *Vorlesungen über Thermodynamik* (Lectures on Thermodynamics) (1897)

Marie Curie "Sur une nouvelle substance fortement radio-active, contenue dans la pechblende" (On a New, Strongly Radioactive Substance Contained in Pitchblende) (1898)

1900

Albert Einstein "On the Electrodynamics of Moving Bodies" (1905)

Werner Heisenberh "Über quantentheoretische Umdeutung kinematischer und mechanischer Beziehungen" (Quantum theoretical re-interpretation of kinematic and mechanical relations) (1925)

1925

Ernest Rutherford *The Electrical Structure of Matter* (1926)

1926

1930

Niels Bohr *Atomic Theory and the Description of Nature* (1934)

1935

Physics

The desire to understand how the natural world around us works and to construct laws and theories that make concrete predictions of how it might act in the future was the genesis of the discipline of physics. The great thinkers in the following chapter went further in doing so, from the very large scale to the subatomic.

Nuclear Science

Releasing even a small proportion of the energy stored inside atoms can have an awesome consequence, but it can also be used for peaceful means, as a source of energy.

The realization that there was a class of chemical elements that spontaneously broke down to release energy and radiation gave rise to a new branch of science that grew in importance through the twentieth century. Before this, scientists had worked out that materials were made of atoms, but had concluded that these atoms were unable to change. Now they could see that under certain circumstances some could change.

By 1934 Italian physicist Enrico Fermi (1901–1954) had shown that neutrons could split many different types of atoms.

But the results he got confused him because he found that adding the mass of the resulting elements gave a result that was much lighter than the original material. In 1938 German scientists Otto Hahn (1879–1968) and Fritz Strassman (1902–1980) produced similar results.

It was a combination of World War II driving all these scientists to work in the United States together with Einstein's famous $E = mc^2$ equation that triggered the next development. Einstein's equation showed that it was possible for mass to disappear, so long as energy was given out. At this point Niels Bohr (1885–1962) also arrived in the United States, and joined an active group of physicists who were discussing the possibility of creating a sustainable chain reaction where the energy released by splitting one atom could be used to split others, and so on.

On the morning of December 2, 1942, this group of scientists, led by Fermi, gathered at a squash court beneath the University of Chicago's athletic stadium. In the court they had built a cubic pile of graphite and uranium that had rods of cadmium running through it. The cadmium was included because it absorbed neutrons. They slowly started pulling the cadmium rods out and monitored the

"There are two possible outcomes: if the result confirms the hypothesis, then you've made a measurement. If the result is contrary to the hypothesis, then you've made a discovery."

–Enrico Fermi

temperature of the stack. At 3:25 pm the stack temperature rose and they allowed the rods to go back in—they had initiated a self-sustaining nuclear reaction and the world had entered the nuclear age.

While the first uses of this technology were military, with the 1945 bombing of the Japanese cities of Hiroshima and Nagasaki, the scientists were quick to turn their ideas into more peaceful applications. On December 20, 1951, in Idaho, the first power station driven by nuclear energy began to produce electricity, and since then nuclear reactors have been built around the world.

The debate now rages as to whether we should make more use of nuclear power in producing electricity. It has a major advantage in that it produces no carbon dioxide or other waste gases that could contribute to climate change, but it has the disadvantage of producing small volumes of waste that will continue to emit radiation for thousands of years. It's a prime example of science generating a technology that seems almost too hot to handle.

Isaac Newton
Understanding Gravity

A formidable mathematician and observer of the physical world, Isaac Newton described universal gravitation, laid the groundwork for classical mechanics, and shares the credit for the development of differential calculus. A genius of the highest order, he is regarded as perhaps the most influential scientist in history.

After performing a series of experiments Newton concluded that two bodies, such as an apple and the Earth, or for that matter the moon and the Earth, attract each other with a force that you can calculate by multiplying the masses of the two bodies, and then dividing that figure by the square of the distance between them. This became known as the Law of Universal Gravitation.

Squaring values subsequently became a frequent aspect of physics equations. As future scientists began to explain gravitational fields the idea of squaring started to make physical sense. Gravity can be explained in terms of lines of flux radiating out from the center of an object; and the more massive an object, the more lines of flux. Think of these as ropes that grab objects and pull them to the center. At any distance from the object, the gravitational effect of these flux lines will be spread over the imaginary sphere that has its center in the center of the object. The greater the distance from the object, the larger the surface area of the sphere, and the more sparsely spread the "ropes."

The area of the surface of the sphere is defined by the equation Area = $4pr^2$, so a change that, say, doubled the distance "r" will quadruple (r^2) the surface area. The gravitation field will then be spread over this much larger area and will be that much weaker.

In his *Mathematical Principles of Natural Philosophy*, Newton extended his ideas and claimed to have identified three of nature's fundamental laws: 1. *That a body at rest, or in uniform motion, will continue in that state unless a force is applied*; 2. *That you can calculate the force applied to an object by measuring the object's weight and the rate at which it accelerates or decelerates*; and 3. *That if one body exerts a force on another, the second body will exert an equal and opposite force on the first.* This Newtonian understanding became the bedrock of physics and continues to have great value.

Born
1643, Woolsthorpe, England

Died
1727, London, England

Newton is famously said to have identified the Law of Universal Gravitation after considering the forces that come into play when watching an apple fall from a tree.

Alessandro Volta
Harnessing the Power of Electricity

At the age of twenty-nine Alessandro Volta started teaching physics at a local high school. Within months of arriving at the school he had built his first invention. Named an electrophorus, this device produced an electric charge from friction in a manner similar to the action of rubbing a party balloon on a sweater.

A few years later Volta became Professor of Physics at Pavia University, where he came into contact with Luigi Galvani (1737–1798), a fellow researcher. One day, while dissecting a frog's leg, Galvani's steel scalpel had touched a brass hook that was holding the leg in place. The leg twitched. Galvani was convinced that this twitch had revealed the effects of what he called "animal electricity"—the life force within the muscles of the frog.

Volta was skeptical and studied whether the electric current could have come from outside the animal. He discovered that bringing two different metals together sometimes caused a small electric current to run, and he correctly guessed that this had occurred when Galvani's scalpel touched the hook.

Taking the idea further, Volta created a column of alternating copper and zinc disks, which he separated with sheets of cardboard soaked in salty water to increase the electrolyte conductivity. This stack produced a constantly flowing electric current, and building stacks of varying numbers of elements produced either more or less powerful currents. His largest column consisted of sixty layers, but he soon found that having more than twenty elements in the stack produced a current that was painful if you held on to wires attached to either end. What Volta didn't know was that all metals hold on to their electrons with different degrees of tenacity. If you place two different metals next to each other, electrons will flow from the one that is relatively more keen to give them up—this is the start of an electrical current.

When Volta demonstrated his stack to the French Academy of Science in 1801, the onlookers were so impressed that Napoleon Bonaparte made him the Count of Lombardy. His contribution to the understanding of electricity was so significant that a key measurement of electricity, the volt, was named after him.

Born
1745, Como, Italy

Died
1827, Como, Italy

When, in 1800, **Volta** experimented with stacking several pairs of copper and zinc disks separated by brine-soaked cloth, he discovered that the metals and chemicals, in contact with each other, produced an electrical current. Volta had created the first electric battery.

Michael Faraday
Understanding Electricity and Magnetism

Born
1791, Newington, England

Died
1867, Hampton Court, England

At the age of fourteen, Faraday started work as an apprentice bookbinder, but enjoyed reading the books more than binding them. In one, he found instructions that enabled him to build his own electrostatic machine. Soon after, he joined the City Philosophical Society.

After attending a Royal Institution lecture given by Humphry Davy (1778–1829), Faraday became Davy's chemical assistant and toured the continent as his valet. Among many luminaries he met on his travels was Alessandro Volta, who inspired Faraday to investigate electricity when he returned to London in 1815.

In 1820 the Danish natural philosopher Hans Christian Oersted (1777–1851) wrote a paper describing how a compass needle deflects from magnetic north when an electric current is switched on or off in a nearby wire. This showed that electricity passing through a wire generated a magnetic field. In 1821 Faraday took this a step further. He pushed a piece of wire through a cork and floated the cork on water. The ends of the wire made contact with blobs of mercury and through these he was able to transmit electricity to the wire. When a magnet was nearby the wire moved each time he applied a current. Bending the wire, he found a way of making it and the cork rotate when he fed electricity through it. He had discovered electromagnetic rotation.

Convinced that energy was always conserved within a system, he decided that if electricity could produce a magnetic field, the reverse should also be true—magnetism should be able to produce electricity. However, it wasn't until ten years later, in 1831, that he showed that moving a powerful magnet near to a coil of wire could cause a brief pulse of electricity to flow in the wire—he had discovered electromagnetic induction.

Faraday quickly found other experimental set-ups that demonstrated induction, including the homopolar generator. In this, a copper disk rotates in a magnetic field, generating an electrical current. Innovators such as de Ferranti (1864–1930) and Tesla (1856–1943) would develop this concept into the first successful generators. Others, such as Ányos Jedlik (1800–1895) and William Sturgeon (1783–1850), saw how induction could be used in reverse, creating the world's first electric motors in the process. The discovery of induction, more than any other, allowed nineteenth-century scientists to turn electricity from a scientific curiosity, powering the world into the twentieth century.

Faraday found that a floating cork attached to a wire could be compelled to rotate when electricity was passed through the wire in the presence of a magnetic field. It was a discovery that eventually led to the invention of the electric motor.

James Clerk Maxwell

Expressing Electromagnetism Mathematically

In the world of physics, the two greats of all time are with little doubt Isaac Newton (1643–1727) and Albert Einstein (1879–1955). Although the two were studying the same phenomena—motion and gravity—the theories they came up with could hardly be more different. One scientist more than any other made this possible, the Scot James Clerk Maxwell.

Maxwell was something of a polymath. At the astonishingly young age of fourteen, he wrote a remarkable paper on the mathematics of ellipses before going on to study at the University of Edinburgh and Trinity College, Cambridge.

In 1856, just a year after graduating from Trinity with a degree in math, he published one of his seminal scientific papers, *On Faraday's Lines of Force*. In this he showed how the electromagnetic field proposed by the scientist Michael Faraday (1791–1867) could be treated mathematically in what have since become known as Maxwell's equations.

A year later, Maxwell turned his hand to astronomy and wrote an award-winning essay that showed that Saturn's famous rings could not be solid, something that was only proven using spectroscopy years later, in 1895.

The years 1861 and 1862 cemented Maxwell's reputation as a giant in the world of physics. It was at this time that he published a famous four-part paper entitled *On Physical Lines of Force*. In it, he was to make what is arguably his most important observation. Not content with having established a mathematical basis for the theory of electromagnetism, he went on in these papers to make a calculation of the speed of propagation of electromagnetic fields. Finding that the speed he obtained was close to the recently established speed of light, he wrote: "We can scarcely avoid the conclusion that light consists in the transverse undulations of the same medium which is the cause of electric and magnetic phenomena." This leap of insight provided the bridge between the work of Newton a few centuries before and Einstein a few decades later.

Maxwell also went on to show that temperature is a result of molecular motion, one of the key planks in the study of thermodynamics to this day.

Born
1831, Edinburgh, Scotland

Died
1879, Cambridge, England

The four equations of electromagnetic theory put forward by **Maxwell** are beautifully simple. Their simplicity hides their importance, something taken by Albert Einstein and transformed into his theory of special relativity.

Wilhelm Röntgen
Discovering X-Rays

Among the problems that Wilhelm Röntgen studied were the curious electrical characteristics of quartz, the influence of pressure on the way that fluids refract light, and the modification of the planes of polarized light by electromagnetic forces. But he is best remembered for discovering the X-ray.

Röntgen studied engineering but soon switched to physics, later becoming Professor of Physics at Würzburg University. One of his experiments involved passing an electric discharge through a chamber containing gas of extremely low pressure to see what would happen. Previous work had shown that doing this with very high voltages could generate a stream of particles that became known as cathode rays. (Such chambers have since been refined to form the cathode ray tubes in conventional television screens, and we now call the particles electrons.)

On the evening of November 8, 1895, Röntgen enclosed the discharge tube in a thick black carton to exclude all light. When he turned off all the lights in the room, a paper plate coated with barium platinocyanide suddenly became fluorescent. He soon found that the radiation causing this was emitted when cathode rays struck the glass end of the tube, and that these rays had a much greater range in air than did the cathode rays. The plate glowed even when it was six feet from the tube.

Intrigued by these rays, Röntgen placed objects of different thickness in their path and found that they exposed photographic plates to varying degrees. Then one day he placed his wife's hand just in front of a photographic plate and shone the rays on it for a short time. When he developed the plate, the result was stunning. There was a ghostly outline of the hand, but the plate also revealed an image of the bones inside her hand and a clear mark caused by a ring she was wearing.

Röntgen went on to discover that the new rays were produced when cathode rays hit a material object. Because the nature of these rays was unknown, he called them X-rays. Later, Max von Laue (1879–1960) and his pupils showed that, like light, X-rays are a form of electromagnetic radiation.

Born
1845, Lennep, Prussia (now Remscheid, Germany)

Died
1923, Munich, Germany

Röntgen was the first scientist to discover the X-ray. Calling it a "Röntgenogram" he gave doctors an unprecedented ability to look inside people's bodies.

Heinrich Hertz
Understanding Electromagnetic Radiation

People who die before the age of thirty-seven don't often leave a huge legacy. But Heinrich Hertz was an exception as a result of his groundbreaking work in the field of electromagnetic radiation. This lasting legacy is revealed every time you tune into your favorite radio station, the unit of frequency being named after him.

Having studied at the Universities of Munich and Berlin, and completed a PhD on electromagnet induction in rotating spheres, Hertz became a lecturer in theoretical physics at the University of Kiel in 1883. Here he studied the recent electromagnetic theory of James Clerk Maxwell (1831–1879). This theory was based on unusual mechanical ideas about the "luminiferous ether"—a hypothetical substance supposed to fill all "empty" space, and which was thought to be the material that allowed light to travel through the universe. Hertz looked at the equations used in this theory and found that you could reconstruct them so that the theory no longer required the ether. Electromagnetic theory had just taken a huge step forward.

Moving to become Professor of Physics at Karlsruhe University in 1885, Hertz soon discovered the photoelectric effect—where ultraviolet radiation knocks electrons from the surface of metal and creates an electric current, which is now the basis of many photovoltaic cells used on items from satellites to road signs.

Three years later, he then generated electric waves using a circuit consisting of a metal rod that had a small gap in it. The gap was small enough for the circuit to be completed by sparks jumping across it, and Hertz showed that these sparks triggered waves of radiation that could be picked up on a second, similar set of apparatus some distance away in the room.

In further experiments he showed that, like light, the waves could be focused or reflected, and that they could pass straight through non-conducting materials. Originally called "Hertzian waves," we now know them as radio waves. Hertz saw no practical use for the discovery, but others were quick to see the relevance. Among them was a young Italian by the name of Guglielmo Marconi (1874–1937) who heard about Hertz's discovery while on holiday in Austria, rushed home and started developing the idea until he could transmit a signal for more than one mile. In 1901 Marconi transmitted a signal across the Atlantic from Cornwall to Newfoundland, and radio came of age.

Born
1857, Hamburg, Germany

Died
1894, Bonn, Germany

Hertz said that the electromagnetic waves he had demonstrated were "of no use whatsoever" and just proved the earlier theories of Maxwell. But he turned out to be wrong when Marconi used them to develop the first radio.

Max Planck
Launching the Quantum Theory

It is frequently the case in science that a deviation from the expected result based on an existing theory leads to a new theory that fits the observed results much better. It was just such a situation that led the German physicist Max Planck to suggest a new way of thinking that gave birth to the field of quantum mechanics.

While studying at the Universities of Munich and Berlin for a doctorate in philosophy, Planck was taught by Gustav Kirchhoff (1824–1887), famous for his work on electrical circuits but also for the conception of black bodies, idealized objects in thermal equilibrium that absorb all the radiation that falls on them. Such objects are a useful model in physics although the construction of a true black body remains in the realms of science fiction.

Many scientists at the time were interested in black bodies. In 1900, the British physicists Lord Rayleigh (1842–1919) and Sir James Jeans (1877–1946) proposed a law based on classical thermodynamics that sought to explain the radiation from a black body at various wavelengths. The problem was that although the law gave very good results for radiation of long wavelengths, it produced nonsense for short wavelength radiation. This problem was later dubbed the "ultraviolet catastrophe."

In the same year, Kirchhoff's former student Max Planck —who shared his professor's fascination with black body radiation—derived a new law based on empirical observations that did not suffer the same catastrophe, agreeing well with results at both short and long wavelengths.

Later that year, Planck reformulated his theory to incorporate statistical mechanics, including the postulate that the energy of electromagnetic radiation was "quantized"—meaning that it could only take a series of allowable values rather than any value from a continuous range. This energy (E), he suggested, was proportional to the wavelength of the radiation (f) via the relationship $E=hf$; h is now known as the Planck constant.

Five years later, Albert Einstein (1879–1955) was to show that quantization of energy was indeed a reality through his explanation of the photoelectric effect. Planck, with the help of Einstein, thus kick-started a branch of physics—quantum theory— that began to challenge many concepts of classical physics.

Born
1858, Kiel, Germany

Died
1947, Göttingen, Germany

Although **Planck** did not immediately recognize the significance of the quantization of energy as an explanation for black-body radiation, the work of other scientists, including Albert Einstein, led to Planck being awarded the Nobel Prize in 1918 for this postulation.

Marie Curie
Discovering Radioactivity

At a time when few women had the opportunity to experience the excitement of scientific research, Marie Curie introduced the world to the marvels of radioactivity. Her groundbreaking work directly led to her death, because she had no way of knowing that radiation emitted from the materials she studied could trigger cancer.

Born Maria Skłodowska, she grew up in Warsaw but moved to Paris, where she registered her name as Marie on arrival. Here she excelled in physics and math and, in 1894, was introduced to Pierre Curie, the Laboratory Chief at the Paris Municipal School of Industrial Physics and Chemistry, whom she married the following year.

Initially Curie researched the magnetic property of steels. In December 1895, when German physicist Wilhelm Röntgen (1845–1923) discovered X-rays and Frenchman Henri Becquerel (1852–1908) found that minerals containing uranium also gave off unknown rays, her focus changed to studying Becquerel's uranium rays. She used a highly sensitive instrument (invented fifteen years earlier by her husband and his brother) that could measure tiny electrical currents that pass through air that has been bombarded with uranium rays. She discovered that the strength of the rays coming from a material depended only on the amount of uranium it contained. In addition, the electrical effects of the uranium rays were unaffected if you pulverized the uranium-containing material, kept it pure, reacted it to form a compound, presented it wet or dry, or exposed it to light or heat. Her conclusion was that the ability to give out rays must be a fundamental feature of uranium's atomic structure.

Curie then discovered that other materials gave off rays and called the phenomenon radioactivity. She also became aware that radioactive materials were often a source of heat and started to speculate about the power potentially locked up in such substances —energy that other scientists would subsequently realize could be released in nuclear power stations and in deadly weapons.

The work of Marie and Pierre Curie, who had isolated naturally occurring radioactive elements, was built upon by their daughter, Irene Joliot-Curie, who—joint with her husband Frédéric—was awarded the Nobel Prize for chemistry in 1935 for the discovery of artificial radioactivity. Both children of the Joliot-Curies, Hélène and Pierre, also became esteemed scientists.

Born
1867, Warsaw, Poland

Died
1934, Sancellemoz, France

Curie already knew, thanks to the work of Röntgen and Becquerel, that her radioactive rays had value in medical imaging. She also showed that they could damage biological tissue, a finding that led to their later use in combating cancer.

Ernest Rutherford
Unlocking the Secrets of Atoms

Ernest Rutherford studied and researched at several key universities around the world, making many important discoveries about the subatomic world during his career. But it was in his native country of New Zealand that he began his work in experimental physics and developed monitoring equipment that could determine whether iron was magnetic at very high frequencies of magnetizing current.

After three failed attempts at getting into medicine, Rutherford succeeded, in 1894, in picking up a grant to study science and found himself working as a research student with Joseph John Thomson (1856–1940) in Cambridge University's Cavendish laboratory. Here Rutherford adapted the detector he had built in New Zealand and used it to investigate some of the properties of insulating materials. Impressed with his ability, Thomson invited him to join a select team studying the electrical conduction of gases.

During this work, Rutherford discovered that there were two distinct forms of rays coming from radioactive elements. Passing a beam of such rays through a magnetic field, he quickly saw that some were bent, while others traveled straight on. The ones that went straight he called alpha particles—which are in fact helium atoms with their electrons stripped off—while those bent by the magnetic field he called beta particles, which turned out to be electrons.

Moving, in 1897, to become a Chair of Physics at McGill University in Montreal, Canada, Rutherford discovered radon, a radioactive gas, and published his first book, *Radioactivity*, in 1904.

It was when he returned to England in 1907, this time to become a Professor of Physics at the University of Manchester, that he had an insight that would change our appreciation of the world. He'd given a student a laboratory practical to run, in which they fired alpha particles at thin films of gold. While most of the particles shot through the gold leaf a few were deflected, and one or two bounced straight back. Rutherford said that this was as if a large naval artillery round had been deflected by a piece of tissue paper.

In 1911 he deduced that this could only have occurred if the mass of the gold atoms was contained in an incredibly tiny nucleus. As a result of this insight, Rutherford became known as the father of nuclear physics, a field that shed new light on the very structure of everything around us in the universe.

Born
1871, Spring Grove, New Zealand

Died
1937, Cambridge, England

The detailed experimental work carried out by **Rutherford** shed new light on the field of radioactivity and helped show that atoms had more detailed structure than previously believed.

Albert Einstein
Developing the Theory of Relativity

Few people have left such a large mark in the public consciousness as Albert Einstein. Born in Germany into a Jewish family, Einstein had little success at school and showed no signs of becoming an international superstar when, in 1901, he took a temporary job as Technical Expert, Third Class in the patent office in Berne, Switzerland.

In 1905, while working at the patent office, Einstein submitted four papers for publication. His papers on Brownian motion, the photoelectric effect, and special relativity are all probably worthy of winning Nobel Prizes, and indeed the Nobel committee did award him the 1921 Prize in Physics for his work on the photoelectric effect. While the photoelectric effect and Brownian motion had given massive support to those claiming that atoms existed, relativity was something entirely new.

At first he introduced the idea of special relativity, in his paper "On the Electrodynamics of Moving Bodies." This theory integrated time, distance, mass, and energy and was consistent with electromagnetism, but omitted the force of gravity. It challenged and overturned Newtonian physics by showing how the speed of light was fixed, and was not relative to the movement of the observer. One of the strengths of special relativity is that it can be derived from only two premises:

1. *The speed of light in a vacuum is a constant (specifically, 299,792,458 meters per second).*
2. *The laws of physics are the same for all observers in inertial frames.*

Despite its simplicity, it had startling outcomes. Special relativity claimed that there is no such thing as absolute concepts of time and size; observers' appreciation of these features were relative, said Einstein, to their own speed.

In 1915 Einstein took the idea further and developed a theory of general relativity. According to this theory, gravity is no longer a force, but a consequence of what he called the curvature of space-time. Unlike special relativity, where reality is different for each observer, general relativity enables all observers to be equal even if they are moving at different speeds. The ideas are mind-bending, and even though parts of them have been challenged, they still remain the grounding for the study of physics.

Born
1879, Ulm, Württemberg [now Germany]

Died
1955, Princeton, New Jersey

Einstein challenged and overturned Newtonian physics in his theory of special relativity; he showed how the speed of light was fixed and, therefore, not relative to the movement of the observer.

Niels Bohr

Increasing Understanding of the Atom

Ask someone to draw an atom and they will most likely sketch the classic "solar system," with a large blob in the middle (the nucleus) and a group of smaller, orbiting dots (a cloud of electrons). Introduced over a hundred years ago by Ernest Rutherford (1871–1937), this model was the most successful of a range of ideas at the time. The only problem was that the model did not work in practice, and it took the Danish physicist Niels Bohr to work out an explanation.

In 1911, Bohr was becoming absorbed in the relatively new field of subatomic physics. Sir J. J. Thomson (1856–1940) had discovered the electron in 1897, while in 1911 Ernest Rutherford's experiments with alpha particles had thrown light on the existence of the atomic nucleus. Bohr was lucky enough to visit both Thomson at Cambridge's Cavendish laboratory in 1911 and Rutherford in Manchester in 1912, learning more of the latter's "solar system" model of the atom.

The problem with the Rutherford model was one of energy. According to classical theories of physics, an electron orbiting a nucleus would gradually lose energy, spiraling in toward it and ultimately crashing into it. Since we are all still here, we can surmise that atoms are inherently stable. The other problem with this model is that the radiation emitted by an electron spiraling to its doom would increase in frequency: this did not agree with observations either.

Bohr's improvement on the Rutherford model was to draw from the work of Max Planck (1858–1947) and suggest that electrons traveled in orbits around the nucleus but could only have specific energies; energy was not continuously lost but only radiated when an electron jumped from one orbit to another.

While Bohr's proposal could have been seen as a clever tweak to hide a problem, his model's success lay in its explanation and also in its agreement with observed physical properties of atoms, specifically emission lines in the atomic spectra of hydrogen atoms that could be explained by electrons jumping between Bohr's allowed orbits. It is little wonder then that the science world heartily embraced Bohr's revolutionary ideas.

Born
1885, Copenhagen, Denmark

Died
1962, Copenhagen, Denmark

Although the revised model of the atom proposed by **Bohr** has now been improved further, it still sits at the heart of the ideas we have about how atoms behave in both the physical and chemical worlds.

Werner Heisenberg
Unraveling Quantum Theory

When Newton looked at forces and movement he saw predictability, and developed explanations for everyday events. When Einstein reinvestigated the same issues he concluded that reality was more complex, but it could be predicted if you took sufficient measurements. When Werner Heisenberg helped develop quantum theory, he took physics into a world that was much less certain.

While still in his early twenties Heisenberg met with the world's greatest names in the scientific community, including Albert Einstein (1879–1955), Niels Bohr (1885–1962), Wolfgang Pauli (1900–1958), and Max Born (1882–1970). Inspired by these pioneers, he subsequently revisited work they had started. Current theories about the atom speculated that electrons orbited a nucleus, much in the way that planets orbit the sun. As he collected data about the way that atoms emit and absorb light he came up with a radically new concept called "quantum mechanics."

They drew together the mathematics of matrices with the physics of wave mechanics. Attracted to the world center of debate, Heisenberg moved to Copenhagen to join Bohr's group of pioneering physicists. Here he spent a lot of time with Erwin Schrödinger (1887–1961) who visited frequently, and who was also actively trying to make sense of this area of physics.

The more Heisenberg studied the mathematics, the more curious he became. He realized that if you knew the position of an electron, you couldn't say anything about its momentum. Conversely, if you detect an electron's momentum, you won't be able to measure its position. In essence, it was always impossible to predict what an electron would do next inside an atom, because of the uncertainty left when you try to measure it.

There are two ways of looking at this. One is to say that the experiments were just not sophisticated enough to do this, but give it a few years and someone would solve the problem. The other was to claim that this was a true reflection of a fundamental property of matter, one he said was described by quantum mechanics. Heisenberg presented this to Wolfgang Pauli in a fourteen-page letter in 1927 and then subsequently presented it to the world. Heisenberg's uncertainty principle had arrived.

Born
1901, Würzburg, Germany

Died
1976, Munich, Germany

The uncertainty principle expressed by **Heisenberg** was described in a thought experiment by Erwin Schrödinger: A cat, a vial of poison, and a lump of radioactive material are placed in a box. The vial will open if one atom in the radioactive material decays. Since no one can predict when this decay will occur, while the box remains closed, the cat is in the "superposition state," of being both dead and alive.

1600

1700

1800

1825

1850

1875

1900

1910

1920

1940

1945

1950

1955

Robert Boyle *The Sceptical Chymist* (1661)

Antoine Lavoisier *Traité élémentaire de chimie* (Elementary Treatise on Chemistry) (1789)

John Dalton *New System of Chemical Philosophy* (1808)

Friedrich Wöhler *Grundriss der Organischen Chemie* (Outlines of Organic Chemistry) (1840)

Dmitry Mendeleyev *The Dependence between the Properties of the Atomic Weights of the Elements* (1869)

Paul Ehrlich *Das Sauerstoffbedürfnis des Organismus* (The Organism's Need for Oxygen) (1887)

Emil Fischer *Introduction to the Preparation of Organic Compounds* (1909)

Linus Pauling *The Nature of the Chemical Bond and the Structure of Molecules and Crystals* (1939)

Dorothy Crowfoot Hodgkin "X-ray Crystallographic Investigation of the Structure of Penicillin" (1949)

Frederick Sanger "The amino-acid sequence in the phenylalanyl chain of insulin" (1951)

Chemistry

The popular image of the scientist is of the white-coated eccentric, tinkering away in a laboratory full of bubbling chemicals. The technology of chemistry has moved on over the centuries but the basic concept is still the same. In this chapter, we read about some of the greats, whose determination and, sometimes, luck has elevated them above their peers.

Chemical Nomenclature

The system for the naming of chemical elements has changed substantially in modern days. One of the earliest scientists to name elements was Antoine Lavoisier (see page 64) who attempted to describe what each element does by their name. Hydrogen, for example, means "water-producing" and comes from its ability to form water when combined with oxygen. Oxygen, another Lavoisier name, means "acid-producing," because he believed it was the essential ingredient of acids—he was ultimately wrong in this but the name stuck. Lavoisier also named carbon, from the Latin for "glowing coal."

Sir William Ramsay was another scientist responsible for naming several elements in the same way, including argon (meaning lazy or not working, as a result of its inertness), xenon (meaning strange), and krypton (from hidden, as it was so rare).

The modern system for naming elements came about following a series of disputes over who first invented or discovered something new. The controversy surrounding the naming of the chemical element with atomic number 105 is a case in point.

Elements with high atomic numbers are inherently unstable and not found in nature. As a result, they can only be "discovered" by creating them artificially in nuclear laboratories.

Two laboratories—the Joint Institute for Nuclear Research at Dubna in the former Soviet Union and the Berkeley lab at the University of California—believed they had discovered it and suggested names. The Soviets wanted it to be called nielsbohrium, after Niels Bohr (see page 54), while the Americans preferred hahnium, after the scientists Otto Hahn.

In an effort to keep the peace, the International Union of Pure and Applied Chemistry (IUPAC)—which is the body that officially sanctions element names—came up with a systematic naming scheme for elements with atomic number greater than 100 that few could argue with.

Each digit in the element's atomic number is assigned a word root according to the following scheme:

"The attempt of Lavoisier to reform chemical nomenclature is premature. One single experiment may destroy the whole filiation of his terms..."

–Thomas Jefferson

0 = nil, 1 = un, 2 = bi, 3 = tri, 4 = quad, 5 = pent, 6 = hex, 7 = sept, 8 = oct, 9 = enn

The three roots are then strung together and the suffix –ium appended. Thus the atomic element 105 has the name unnilpentium. The chemical symbol is just the initial letters of the three roots, viz Unp.

In the end, unnilpentium got itself a so-called trivial name after all. In 1997, IUPAC ruled that the Soviet (by this time Russian) team had scientific priority. Rather than calling it by the previously disputed name of nielsbohrium IUPAC suggested dubnium after the location of the Russian lab.

The most recent names granted respectability by IUPAC are flerovium (Fl, 114) and livermorium (Lv, 116), both named after the nuclear laboratories where they were discovered: the Flerov Laboratory for Nuclear Reactions and the Lawrence Livermore National Laboratory.

Robert Boyle
Understanding the Nature of Gases

Born
1627, Lismore, Ireland

Died
1691, London, England

Robert Boyle developed an early passion for alchemy, a subject that, in the seventeenth century, was studied by a highly secretive network of characters who searched for ways of turning base materials into silver and gold. This led to him undertaking a wide range of experiments on the subject of matter.

In 1661, Boyle broke with the alchemists' obsession with secrecy when he published the groundbreaking work *The Sceptical Chemist*, in which he presented the idea that matter consisted of atoms and clusters of atoms. He suggested that these atoms moved around and collided with each other and that these collisions may cause new clusters, with new properties. Crucially, he argued that the atoms remained unchanged, and that, under the right conditions, you could take these newly created compounds and split them back into their original elements.

Boyle had also been carrying out experiments with Robert Hooke (1635–1703). These experiments focused on the properties of air and were made possible because Hooke had developed a sophisticated air pump. The pair had placed a lighted candle under a bell-jar, then pumped out the air. The flame was extinguished as a result. A burning coal in the airless bell-jar ceased to glow, but re-ignited if the air was returned before the coal cooled down. Clearly air was needed for items to burn. Using the same equipment, Boyle and Hooke found that air was also important for the transmission of sound: they managed to put a bell inside the jar, pump out the air, and then strike the bell. With no air in the jar, they couldn't hear the bell.

Consequently, Boyle concluded that there was an intriguing relationship between volume and pressure but he realized that to make specific discoveries you needed to change one thing at a time. In studying gas he identified four variables that needed considering: the amount of gas, its temperature, its pressure, and the volume of the container that is holding it. During his investigations, Boyle therefore fixed the amount of gas and its temperature, but varied the volume of the container or the pressure exerted on it. When he halved the container's volume, the pressure of the gas doubled. If he decreased the surrounding pressure, the container's volume increased. We now know this relationship as Boyle's Law—for a fixed mass of gas, pressure and volume are inversely proportional. It formed the basis of all future work on the physical properties of gases.

Boyle is best known for his law that describes the inversely proportional relationship between the pressure and volume of a fixed amount of gas if the temperature is kept constant within a closed system.

Antoine Lavoisier
Naming the Gases

In an attempt to explain fire, a substance known as phlogiston was conjured up by scientists in the mid 1600s. This combustible component was believed to be contained within all substances and objects that could burn. It took eighteenth-century French aristocrat Antoine-Laurent de Lavoisier to show that phlogiston was a flight of fancy and that it was the presence of oxygen that caused things to burn.

Educated at the French capital's prestigious Collège des Quatre-Nations, Lavoisier studied a wide range of subjects including astronomy, chemistry, mathematics, and botany. By the age of twenty-one, he had trained to become a lawyer but his career changed direction when he followed his passion for chemistry, which had become a hugely popular subject at the time because of the discoveries of the likes of chemist Joseph Priestley (1733–1804), who identified a gas that he called "dephlogisticated" air, which Lavoisier would later name as oxygen.

Experimenting with combustion in an effort to discredit the phlogiston theory, Lavoisier created an experimental set-up in which objects could be burned and the resultant products weighed. In 1777, he published a paper entitled *Memoir on Combustion in General* based on these experiments. He noted that combustible items would only burn in a particular type of air—the "dephlogisticated" air of Joseph Priestley—and that the fire would soon extinguish if placed in a vacuum.

Crucially, he also noted that the burning object increases in weight exactly in proportion to the quantity of pure air destroyed or decomposed—a hugely important principle of chemistry, that of mass conservation. Accordingly, he stated, "the existence of the matter of fire, of phlogiston in metals, sulfur, etc., is then actually nothing but a hypothesis."

He was also interested in the formation of acids and, in 1778, renamed Priestley's dephlogisticated air oxygen—a name taken from Greek, meaning acid-forming and based on his view, later proven erroneous, that it was a necessary component of all acids. His naming of elements did not stop there. He named hydrogen in 1783 and also coined names for more than thirty other elements and substances, most of which are still in use today.

On Lavoisier's death (he was beheaded during the French Revolution), the mathematician Joseph-Louis Lagrange (1736–1813) remarked: "It only took a moment to let his head fall; a hundred years may not be enough to make another like it."

Born
1743, Paris, France

Died
1794, Paris, France

Lavoisier demonstrated how water could be decomposed into its two constituent gases, hydrogen and oxygen. The chemical formula H_2O may be known to every child today but until then, the make-up of water had been unknown.

John Dalton
Explaining Chemical Reactions

It is hard to imagine a world without the familiar concept of distinct elements made up of atoms with particular chemical properties but it was not until a young Englishman called John Dalton interested in the weather turned his mind to creating an atomic theory of chemistry that the picture we know today emerged.

Born to a Quaker family, Dalton was a teacher for most of his life—in the fields of mathematics and natural philosophy. He had developed an early fascination with math and meteorology under the guidance of Quaker meteorologist and instrument maker Elihu Robinson. Dalton's interest in the gases of the atmosphere led him to propose the law of partial pressures, now known as Dalton's Law. This states that the pressure exerted by each gas in a mixture is independent of the pressure exerted by the other gases, and that the total pressure is the sum of the pressures of each gas. He soon began to believe that the atoms of each of the gases had a fundamental nature that explained their chemical interactions.

This proposal of an atomic theory of chemistry was inspired in part by Antoine Lavoisier's law of mass conservation (see page 64) and Joseph Louis Proust's law of definite proportions, which states that when a compound is broken down into its constituents, the weights of those constituents are always in the same ratio, no matter how much of the original substance there is.

In 1803, Dalton published a table showing the relative atomic weights of six elements—hydrogen, oxygen, nitrogen, carbon, sulfur, and phosphorus. He took as his base the element hydrogen, the lightest known, which was assigned a value of one. Rather than theorizing about the inner subatomic structure, which would not be clear for another century, Dalton based his ideas on empirical observations, which is probably what drove him to list the atomic weight of oxygen as seven rather than eight as we know it today.

In a lecture to the Royal Institution of Great Britain the same year, Dalton went further and proposed a fully fledged atomic theory based on a number of key principles: all matter is composed of atoms, which cannot be made or destroyed; all atoms of the same element are identical; different elements have different types of atoms; chemical reactions occur when atoms are rearranged; and compounds are formed from atoms of the constituent elements.

Dalton incorrectly believed that a water molecule consisted of one hydrogen atom and one oxygen atom, or HO rather than the now familiar H_2O. Despite this, his theories stand the test of time.

Born
1766, Eaglesfield, England

Died
1844, Manchester, England

By categorizing elements according to their relative weights, **Dalton** gave an early hint as to the structure of the atomic nucleus. He was unaware of protons and neutrons but his table helped to establish the idea of elements with specific chemical properties.

Friedrich Wöhler

Synthesizing Organic Compounds

Born
1800, Eschersheim, Germany

Died
1882, Göttingen, Germany

It is a tradition of science that before experimenting on others, it is the decent thing to carry out experiments on oneself. Many medical developments would never have happened without this simple precept. Friedrich Wöhler was one scientist to do this, and his paper on how waste products pass into urine, using both his own and that of his dogs as examples, is considered a classic of the genre.

Wöhler seemed destined for a career as a doctor, studying medicine at Heidelberg University and obtaining an MD in 1823. But while there he attended lectures by the chemist Leopold Gmelin (1788–1853) and was, in turn, encouraged to take up chemistry, for which he had a great aptitude. Gmelin was instrumental in getting Wöhler to spend a year with Swedish chemist Jöns Jakob Berzelius (1779–1848) in Stockholm, where Wöhler became a great experimentalist.

Wöhler's greatest contribution to science, made when he returned from Sweden to Germany, is best put in the words that he himself wrote: "I can no longer, as it were, hold back my chemical urine; and I have to let out that I can make urea without needing a kidney, whether of man or dog." What he meant by this was that he had worked out how to synthesize urea by reacting lead cyanate and ammonium hydroxide with each other. It was the first time that someone had managed to create a natural product outside a living organism. This discovery also demonstrated the concept of isomerism—the idea that two distinct compounds could have the same chemical formula.

Wöhler also made great advances in the isolation of several metals, notably boron, silicon, beryllium, titanium, and, particularly, aluminum. Many great chemists, including his mentor Berzelius and the great Humphrey Davy (1778–1829), had tried to isolate aluminum without success. Wöhler's treatment of anhydrous aluminum chloride with a potassium amalgam therefore succeeded where these others had failed.

Yet it was his synthesis of urea that was to secure his place in history. The discovery of this creation process is considered so vital that it is often said to be the moment that the discipline of organic chemistry—the chemistry of carbon-containing substances—began.

The synthesis of an organic compound outside a living body by **Wöhler** was the nail in the coffin for the theory of vitalism, which held that living processes were somehow unique to living organisms.

Dmitry Mendeleyev
Classifying the Chemical Elements

Born in the Siberian town of Tobolsk, Mendeleyev made his most important contribution to chemistry when he was thirty-five, teaching at St. Petersburg University. By this time, chemistry had moved a long way forward from the days of the seventeenth-century alchemists who believed that materials could be transformed from one to another.

Now chemists knew that at a chemical level, materials were built of unchanging elements, and that these elements combined to make molecules. But one question started recurring in many chemists' minds: Why did different materials sometimes look and behave alike? Based at the University of St. Petersburg, Mendeleyev therefore set about looking at similarities in the behaviors of different elements.

According to his notes, his periodic table came as a spark of inspiration while he was setting out to write a new chemistry textbook, *The Principles of Chemistry*. In a remarkably creative few hours on February 17, 1869, Mendeleyev sat down with sixty-three cards. On each card he wrote the name of one element, its atomic weight, and physical and chemical properties. This pack contained all of the elements known at that time. By sorting the cards in a gridlike pattern, Mendeleyev realized that you could place them so that the atomic weight increased as you went down a column, but elements in any row shared similar properties. The first column started with lithium, followed by beryllium, boron, carbon, nitrogen, oxygen, and fluorine.

In a modern periodic table the grid has been turned sideways and this set of elements appears as the first row. Equally he grouped some elements according to their properties, even though the weights didn't fit the pattern. He rightly assumed that some elements were yet to be discovered and that the recorded weights of some were wrong. Over time, as the missing elements were found and the weights were recorded accurately, chemists discovered that they matched Mendeleyev's predictions. Designing a system for explaining what you know is clever; using that explanation to make accurate predictions is remarkable.

The Periodic Table remains to this day the roadmap that allows chemists and physicists to explore the world of elements and lets them characterize their properties with confidence.

Born
1834, Tobolsk, Siberia, Russia

Died
1907, St. Petersburg, Russia

Mendeleyev devised the Periodic Table, in which the elements are arranged in groups (columns) and periods (rows) according to their properties. Cleverly, he left gaps in the table, guessing—correctly—that some elements were yet to be discovered.

Emil Fischer

Unlocking the Chemistry of Sugar

Born
1852, Euskirchen, Germany

Died
1919, Berlin, Germany

Emil Fischer was a brilliant chemist, correctly detailing the structure of many key organic substances before going on to create many of these in the lab for the first time ever. He is best known for his work on the sugars, such as fructose and glucose.

Originally destined to follow his father into the lumber trade, Emil Fischer was instead convinced by his cousin Otto to study chemistry, first at the University of Bonn and then at Strasbourg under the eminent professor Adolf von Baeyer (1835–1917). Fischer worked with von Baeyer and eventually synthesized a substance called phenylhydrazine, used in the dye and pharmaceutical industries, which set him on a bright career in organic chemistry.

However, it was at the University of Erlangen, where he moved to in 1881, that he did his most famous work. He began by analyzing some of the chief organic compounds in tea and coffee, particularly caffeine and theobromine, and worked out how to produce them synthetically.

Extending his work to compounds found in animal excreta, including uric acid and guanine, as well as adenine, he began to realize that they all—including caffeine—belonged to a single family of compounds, which he called the purines.

In 1885, Fischer left von Baeyer's laboratory for the University of Würzburg where he changed the direction of his research for the next decade to concentrate on the organic chemistry of sugars. He worked in the same methodical way as before, watching how these compounds reacted and how they degraded before suggesting their stereochemistry, or three-dimensional structure. (In so doing, he developed the Fischer projection method, still used today, which allows chemists to represent three-dimensional shapes on a sheet of paper.)

He then moved on to synthesizing the compounds. In 1890, he suggested how the sugars glucose and fructose are related and succeeded in synthesizing both—an important breakthrough in that it was the first step in the artificial production of complex sugars that are rare in nature.

Fischer also did much pioneering work on proteins and enzymes, suggesting the method of operation of the latter was analogous to a lock and key. This analogy has been vital in understanding how to develop enzymes that can treat disease or manipulate genes. Fischer's own work was like one of these enzymes, unlocking the mysteries of organic chemistry.

As part of his organic chemistry research, **Fischer** analyzed amino acids and uncovered the nature of the bonds that link them together to create proteins—a vital breakthrough in biochemistry.

Paul Ehrlich

Applying Chemistry to Medicine

Paul Ehrlich was hugely influential in the development of what are now known as "blockbuster" drugs. His concept of "magic bullets"—his term for synthetic chemicals—binding to receptors of a target disease and destroying it remains central to the pharmaceutical industry and the way illnesses are treated today.

Ehrlich's work in this field began with his research into the use of dyes for staining tissue specimens. In 1882 he published a method of staining the tuberculosis bacillus that is still used today, albeit in modified form. In 1899, Ehrlich became Director of the newly established Royal Institute of Experimental Therapy in Frankfurt. He turned his attention to chemotherapy—specifically how the chemical structure of drugs related to their mode of action in the cell and their affinity for cell components. He formulated his "side chain theory," explaining how specific dyes bound chemically to specific locations within the cell (side chains) and how antibodies, similarly, could bind to antigens. In 1900, he and his assistant, the bacteriologist Julius Morgenroth, renamed the side chains "receptors." Ehrlich began researching synthetic chemicals (his magic bullets) that would bind to receptors on a disease target—whether a cancer cell, inflammatory molecule, virus, or bacterium—and so destroy (or at least neutralize) it.

He began testing hundreds of substances for magic-bullet properties and found a dye, Trypan Red, which was active against trypanosome parasites. He began work on syphilis, which was caused by the newly discovered spirochaete bacterium *Treponema pallidum* and tested many different arsenic compounds. His assistant, Sahachiro Hata, discovered that compound number 606 worked against syphilis in rabbits and, later, in humans. Its introduction, in 1910, under the name Salvarsan was the first treatment for syphilis, and in 1912, an improved version, Neosalvarsan was introduced.

Chemotherapy did not really gain acceptance until the large-scale production of penicillin during World War II, when it was shown to be very effective against syphilis. Since then, chemotherapy has become widely known, although the term is now almost exclusively reserved for its use in the treatment of various forms of cancer. The concept of fighting disease with targeted drugs, however, is now virtually universal.

Born
1854, Strehlen, Germany

Died
1915, Bad Homberg, Germany

Ehrlich found that dyes allowed for more detailed work to be done under the microscope and classified his dyes into three groups: basic, acid, or neutral. His work on staining structures within blood cells formed the basis of modern hematology.

Linus Pauling
Explaining the Carbon Bond

Winning a Nobel Prize is surely the pinnacle of any scientist's career. Given the nature of scientific research, the likelihood of winning one unshared with others is becoming increasingly remote. This makes Linus Pauling's achievement of winning two unshared Nobel Prizes —one for chemistry and one for peace—even more remarkable.

Pauling's achievement is less surprising when you consider his wide-ranging interests: He published hundreds of scientific papers on a bewildering range of subjects, from X-ray diffraction and the structure of metals, to proteins, antibodies, and hemoglobin.

However, it is as a result of his work on the nature of chemical bonds that he has secured his place in the history of science. His insight was to combine the then new theories of quantum mechanics with the shared electron theory of chemical binding. In particular, in 1932, he suggested how quantum mechanics could explain the challenge posed by carbon; physicists said it should form two strong bonds while chemists knew that it made four, making the classic tetrahedron shape. Pauling's breakthrough came when he worked out how to use the wave functions of quantum mechanics to describe the probabilities of where these four electrons would be in their orbits; his calculations showed that carbon should have a natural tetrahedral shape.

In 1939, Pauling published what was to become one of the most influential chemistry books of all time: *The Nature of the Chemical Bond and the Structure of Molecules and Crystals*. Although this book dealt with a weighty subject—how quantum mechanics governed the behavior of molecules and compounds—it was written in a way that was accessible to undergraduates, leading to it becoming a standard textbook.

In 1954, he was awarded the Nobel Prize for chemistry. It was then Pauling's unstinting opposition to the bombs that had ended World War II and initiated the Cold War that won him his second Nobel Prize eight years later.

Pauling is also at least partly responsible for the notion that vitamin C is effective in preventing and reducing the severity of the common cold and of preventing cancer, although its efficacy is strongly debated even today.

Born
1901, Portland, Oregon

Died
1994, Big Sur, California

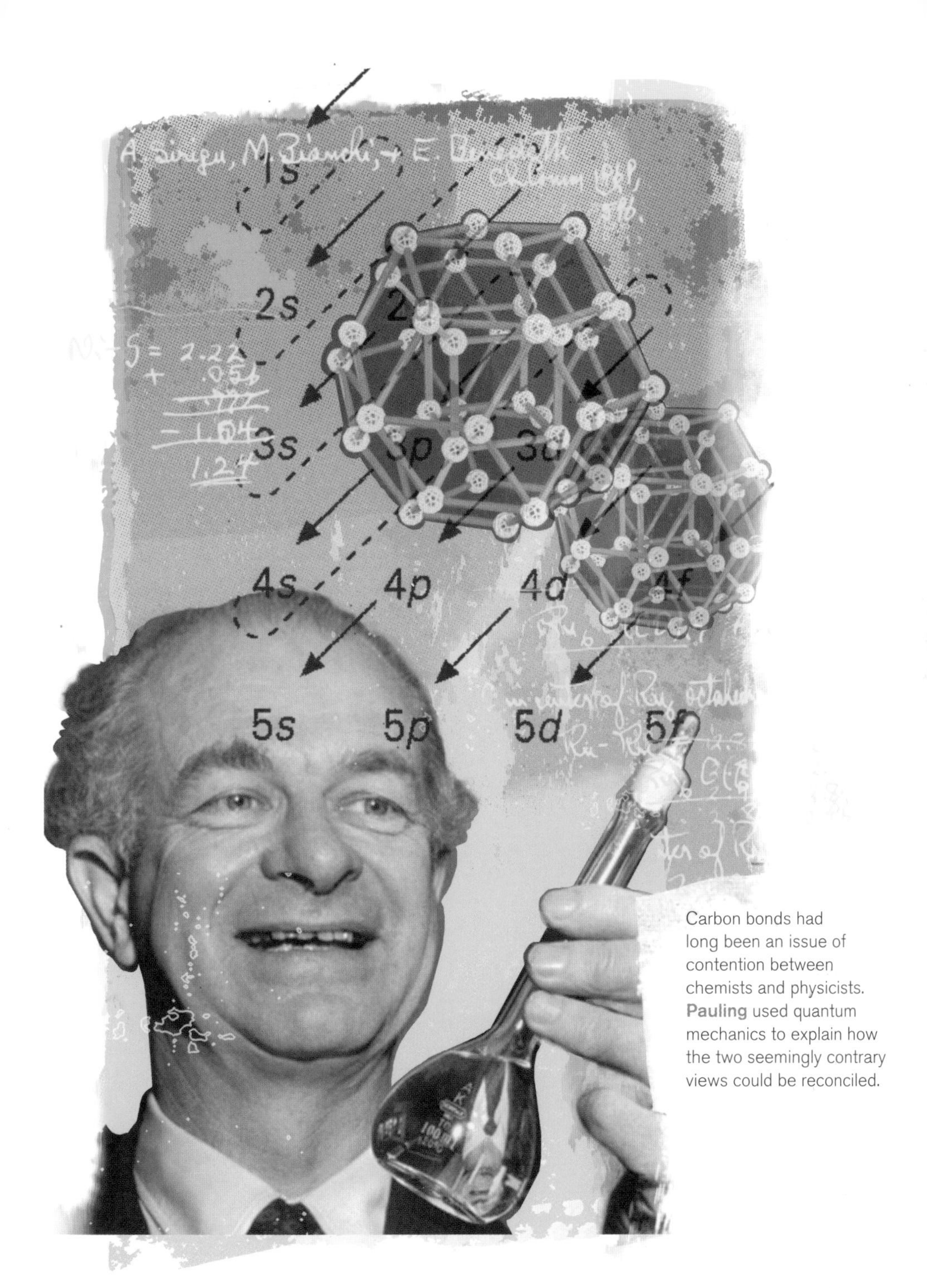

Carbon bonds had long been an issue of contention between chemists and physicists. **Pauling** used quantum mechanics to explain how the two seemingly contrary views could be reconciled.

Dorothy Crowfoot Hodgkin

Observing the Structure of Molecules

In the latter part of the nineteenth century, chemists started calculating the shapes of some of the large, carbon-containing compounds that are fundamental to life. They made assumptions about what bonds would form and built scale models to describe the compound structures. Although the exercise worked reasonably well with small molecules, it failed with the larger ones.

In the first years of the twentieth century scientists began to realize that when they shone X-rays through crystals and onto photosensitive paper, they got patterns. The idea is that when X-rays hit a crystal, the electrons surrounding each atom bend the beam. Because there are many atoms arranged in repeating patterns within the crystal, the X-rays produce a series of light and dark patches. Measuring the intensity and relative position of each patch gives clues about the relative positions of atoms within the crystal. Now people could start to make sense of the three-dimensional structure of compounds.

Born in Cairo, Egypt, and educated at Oxford University, Dorothy Crowfoot Hodgkin started to develop techniques of X-ray crystallography. One of her early successes was making predictions about the structure of a small protein called pepsin. Hodgkin then started work on insulin, a task that took her thirty-four years to complete.

In the 1930s and 1940s the germ-busting antibiotic penicillin arrived on the scene. Initial clinical hopes for this drug were dashed because it was difficult to harvest from microbes, and there was no hope of manufacturing it until someone could work out its structure. Hodgkin turned her X-rays on it and found that it had an unusual ring feature, now known as the Beta-lactam structure. This discovery started to give important clues about how the antibiotic worked.

With this pioneering work in X-ray crystallography, Hodgkin was able to discover the chemical structure of penicillin and insulin, which enabled them to be manufactured synthetically and become widely available to those in need. She demonstrated that we could look inside molecules and see their structure. From here it is a short step to predict their function and work out how to design chemical drugs that could affect them. Biochemistry and the pharmaceutical industry would never be the same again.

Born
1910, Cairo, Egypt

Died
1994, Shipston-on-Stour, England

Hodgkin showed how calculating the three-dimensional structure of molecules, allowed them to be recreated synthetically by the pharmaceutical industry.

Frederick Sanger

Deciphering the Language of DNA

Winning one Nobel Prize is a rare feat, but to receive two is truly remarkable. Frederick Sanger got his first for showing how amino acids link together to form the protein insulin, and his second for developing a method of sorting out the sequence of molecular "letters" that make up a genetic code.

Studying at the University of Cambridge, Sanger gravitated to biochemistry. Joining a team looking at the structure of proteins, he turned his attention to insulin. At the start of his work it was possible to look at the protein using an electron microscope and see its overall shape, or to mash it up chemically into a soup of twenty-two building blocks called "amino acids." Scientists knew that these amino acids were normally linked together in a long chain, but had no clue as to their sequence in that chain.

Sanger marked the end of the protein chain with a dye that stuck even when the chain was dismantled. Breaking the chain into lengths of two, three, four, five, or more amino acids, he identified the amino acid at the end of each fragment and, by doing this enough times, in 1951 he was able to determine the order of fifty-one amino acids in a molecule of insulin. The same method could be used to analyze any other protein of interest.

When it came to working on another chainlike structure, deoxyribonucleic acid—DNA, Sanger wanted to determine the sequence of its "bases," or building blocks. By this time scientists knew that it was this sequence that held the genetic instructions that make up our bodies.

DNA code is written using A, C, T, and G, in sequences of three letters at a time (TGC or TGG, for example). Each sequence is called a DNA triplet or codon. Most groups code for an amino acid, while those that don't provide the grammar of the DNA sequence. For instance, the codon TAA means "stop," essential for telling the cell when to halt producing proteins.

Sanger generated fragments of the sequence, this time ending at each different base. By measuring the length of each fragment he showed that you could determine the complete sequence. The process lends itself to computer automation, which has allowed scientists to tackle huge sequences, including the three-billion-base sequence that makes up the human genome—the DNA inside the nucleus of each human cell.

Born
1918, Rendcomb, England

Died
2013, Cambridge, England

Sanger showed how amino acids link to form proteins, particularly through his work on sequencing the structure of insulin, and how to read genetic code, using the DNA triplets or codons to help determine the sequence of its bases.

Claudius Galen *De libris propriis* (On My Own Works) (2nd century CE)

Avicenna Kitab al-Shifa' (The Book of Healing) (1020)

Andreas Vesalius *De Humani Corporis Fabrica* (On the Fabric of the Human Body) (1543)

William Harvey "De Motu Cordis" (On the Motion of the Heart and Blood) (1628)

Carl Linnaeus *Systema Naturae* (1735)

Claude Bernard *Introduction à l'étude de la médecine expérimentale* (Introduction to the Study of Experimental Medicine) (1865)

Karl Landsteiner *Die Spezifität der serologischen Reaktionen* (The Specificity of Serological Reactions) (1936)

Carl Djerassi *Optical Rotatory Dispersion* (1960)

Biological Systems

Biology is the broadest of the sciences, covering everything from the simplest of plants to the most complicated of organisms. Thanks to the pioneering biologists in this chapter, from ancient Greece to modern day, we have made normous breakthroughs in understanding the plant and animal kingdoms.

Anatomy

The study of human anatomy has a complex history, but over time anatomists have come to gain a significant insight into how the body works. This has radically affected the way doctors and surgeons try to cure disease or treat injury.

Whether you are comfortable with the idea of poking around inside dead bodies depends a lot on your view of the nature of life. Throughout history, some societies have seen it as a horrific intrusion into a person's being, while others have used it as the ultimate punishment for convicted murderers. Still others have looked inside bodies either as a means of fortune telling, or of studying how the human body works.

The earliest records of anatomical studies are found on fragments of clay tablets made around 4000 BCE in Nineveh, an ancient city in what is now northern Iraq. It appears that the temple priests made clay models of organs, such as the liver and lungs of sheep, and used observations of these organs for some form of fortune telling.

A more developed understanding of anatomy can be seen in papyrus fragments written in about 1000 BCE and found in Egypt. These show a relatively complex understanding of features such as our eyes, the digestive system, and bones. Much of this information probably came from the practice of mummifying bodies before burial.

According to Tertullian, a historian who lived in the ancient city of Carthage, in what is now Tunisia, sometime around 200 CE, the study of human anatomy started to take off in Alexandria, Egypt. He says that Herophilus (ca. 335 BCE–ca. 280 BCE) and Erasistratus (born ca. 250 BCE) vivisected up to 600 criminals. The Roman scholar Celsus (ca. 30 BCE–45 CE) says that Herophilus obtained these criminals "for dissection alive, and contemplated, even while they breathed, those parts which nature had before concealed." All this work was lost, presumably in the great fire that destroyed the library of Alexandria in the early years of the Christian era.

"(The heart) is the household divinity which, discharging its function, nourishes, cherishes, quickens the whole body, and is indeed the foundation of life, the source of all action."

–Willam Harvey

A major step forward occurred when the Greek physician Claudius Galen (ca. 129 CE–ca. 200 CE) started studying anatomy, getting much of his insight while working with wounded gladiators in Rome. However, between Galen's observations and the late fifteenth century, very little new anatomical work was done. People felt that dissection was unethical and offensive on religious grounds, and Galen's work was simply taken as true and translated into many different languages. Then in 1490 the world-leading medical school at Padua in Italy opened a new anatomical theater. This stimulated people such as Leonardo da Vinci (1452–1519) and Andreas Vesalius (1514–1564) to start dissecting bodies again in an attempt to work out exactly what went on inside.

The seventeenth and eighteenth centuries saw the dawn of the Enlightenment in Europe, a golden era for science in which philosophers and scientists started to question previously held beliefs that had seen everything associated with life in terms of purely spiritual forces. Instead they looked for mechanical, verifiable explanations. Studying anatomy then became part of the process of seeing how the human machine functioned, a process that has continued ever since.

Claudius Galen
Understanding Anatomy

Born
Ca. 129 CE, Pergamum, Mysia [now Bergama, Turkey]

Died
Ca. 200 CE, place unknown

As the Christian era dawned, the human body was considered predominantly as a vessel inhabited by a spirit. The vessel itself was of little interest—it was the spirit that demanded attention. Galen disagreed. He was fascinated by the physical structure of the body, and wanted to work out what each organ did.

Born to wealthy parents living in Pergamum, a thriving city on what is now the west coast of Turkey, Galen started studying medicine at the age of sixteen. At twenty he headed off to Smyrna in Greece to study anatomy with the respected physician Pelops. In his early thirties, Galen moved to Rome.

The Roman Empire's love of blood sports, especially those involving human combat, gave Galen an extraordinary access to subjects. By being on hand to witness the effects of serious injury, he realized that two different types of blood came from deep cuts. One type was dark blue and ran sluggishly from cut vessels. These vessels had thin walls and, by dissecting corpses, Galen found that he could trace these vessels back to the liver. As the liver was intricately connected to the gut, he concluded that food was broken down in the gut to form chyle, which was passed to the liver and there turned into blood. This then flowed out to the other organs of the body, where it was used up.

Galen also noted that one vessel went from the liver to the heart. He was convinced that there were pores that allowed the blood to move between the chambers of the heart. His observations of animals as well as dead and dying soldiers revealed that the blood leaving the heart was bright red and seemed to be packed with life. It spurted from the vessel when it was cut and, if allowed to flow, a person's life soon left his body. Galen's conclusion was that the heart packed the blood with vital spirit, the stuff of life, and that this pulsating fluid then distributed life throughout the body. In many ways these were good observations, but his conclusions were wrong and it took 1,700 years before William Harvey gained a better insight into blood and how it circulates around the body.

Galen made a large number of key anatomical observations, many of which related to blood, thanks to his first-hand contact with soldiers injured in blood sports.

Avicenna
Documenting Medical Knowledge

Avicenna developed a system of medicine that is famous for synthesizing elements of Eastern and Western knowledge, including Galen's work, Islamic medicine, Aristotelian philosophy, and a range of concepts from traditional Indian medicine.

Avicenna is the European name of Ibn Sina, who lived in the so-called Golden Age of Islam (ca. 700–1200), when knowledge of mathematics, philosophy, and medicine blossomed. A prolific writer, he is credited with having produced as many as 450 works, of which 240 survive. Of these, forty concern medicine while others are on music, geometry, astronomy, religion, and philosophy.

His most noted medical work is his fourteen-volume "Canon of Medicine," which was translated into Latin by Gerard of Cremona in the twelfth century and subsequently introduced into Europe. The Canon consists of five different books and one million words covering general principles; pharmacology; diseases specific to a given organ; non-specific diseases, like fever; and recipes for remedies. The pharmacology work lists 760 drugs with comments on their effectiveness.

The Arabic text of the Canon was printed in Rome in 1593, making it one of the first Arabic texts in print. Owing to its encyclopedic content and its structured layout, it soon replaced the works of Galen in its importance to the growing medical profession and remained a key volume up until the eighteenth century. The Canon continued as a medical text in the Islamic world long after it had been superseded in Europe, and is still valued today in certain medical schools in India and Pakistan.

Avicenna introduced many medical concepts that are still familiar today, such as risk factors; the importance of diet, climate, and environment in health; clinical trials; contagion; quarantine; and experimental medicine. He was the first to describe the anatomy of the human eye and eye diseases such as cataract. He also realized that tuberculosis was contagious and that diseases can be spread through soil and water.

Furthermore, Avicenna was the first to take a clinical interest in psychology and psychiatry and is credited, through his interest in chemistry, as an early pioneer of aromatherapy.

Born
980, Afshana, Central Asia

Died
1037, Hamedan, Persia

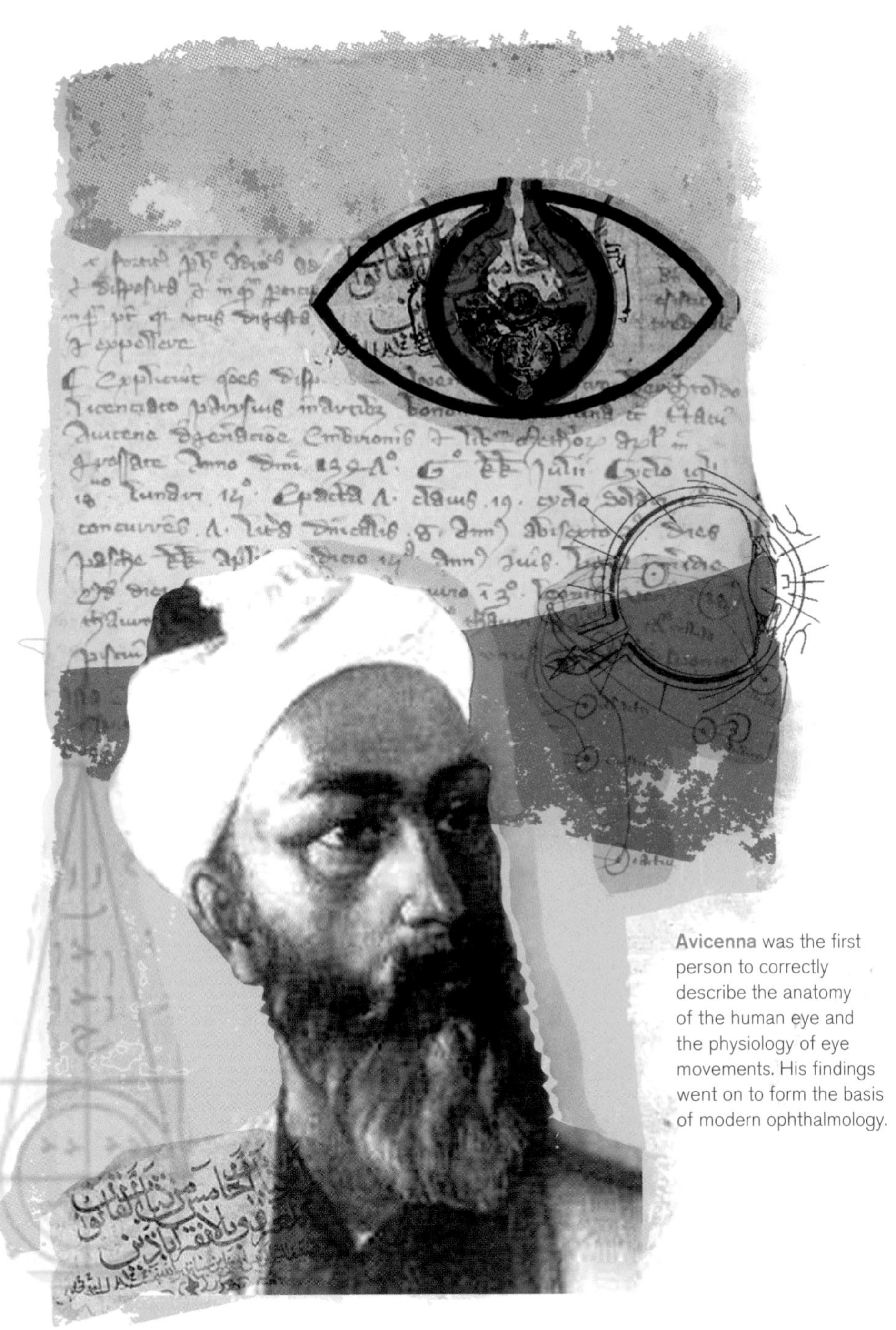

Avicenna was the first person to correctly describe the anatomy of the human eye and the physiology of eye movements. His findings went on to form the basis of modern ophthalmology.

Andreas Vesalius
Championing Dissection

Andreas Vesalius produced a milestone in medical science when he published *On the Structure of the Human Body.* Based on research using human dissection, the book gave precise descriptions and illustrations of the human skeleton, muscles, nervous system, blood vessels, and organs. Controversial at the time of publication, the work challenged Galen's theories on anatomy, which had dominated for well over a thousand years.

The son of a Brussels pharmacist, Vesalius studied medicine at Louvain in Belgium in 1528, moving to Paris in 1533. He returned to Louvain in 1536, as a result of war in France, and subsequently studied in Padua. It was common for doctors to study in several leading centers of knowledge during the Renaissance period. After a short period as a military surgeon, he took up a professorship in anatomy and surgery at the University of Padua, at the age of twenty-four.

Vesalius preferred to carry out human dissections himself, rather than leave the work to an assistant, and used this as a tool for research into human anatomy. Until this time, dissection had been used mainly for instruction (even this was controversial, as it went against many religious traditions to cut into the human body). His interest increased when a Paduan judge allowed him to work with the bodies of executed criminals.

Vesalius criticized Galen's medicine, which had until this time held sway in the medical profession. For example, Vesalius challenged Galen's claim that there is a passage between the two ventricles of the heart through which blood can pass.

Initially, Vesalius's work met with considerable resistance in Padua and, subsequently, in Spain, where he became physician to King Charles V and then, in 1555, to his son, Philip II. However, by the end of the sixteenth century, the view of human anatomy presented by Vesalius had become the "gold standard" in medicine and was widely translated and disseminated. Vesalius had laid the foundations for more detailed studies of human organs, his accurate understanding of human anatomy proving essential to William Harvey's later work on the circulation of the blood and to other work promoting the view of the human body as an efficient machine.

Born
1514, Brussels, Belgium

Died
1564, Zakynthos, Greece

Vesalius was one of the first anatomists to collaborate with an artist trained in anatomy. In *On the Structure of the Human Body*, woodcut illustrations—probably produced by German artist, Jan van Calcar—represent the dissected human body in naturalistic poses.

William Harvey
Discovering Blood Circulation

Up to the seventeenth century, philosophers had ruled the world of thought with the idea that you could deduce how something worked by thinking about it and seeing how your conclusions fit with the foundational work of great ancients, such as Galen and Aristotle. Prior to William Harvey, carrying out experiments and assessing the data obtained were revolutionary concepts.

While Harvey had a great respect for the ancients, his sense of curiosity drove him to investigate for himself. Having been trained in Padua, Italy, at the world's foremost medical school, he returned to England. By nurturing an association with King Charles I, Harvey got permission to use the king's deer and other animals, which gave him the rare ability to perform experiments on a wide range of large, live animals. He found that performing experiments could turn established ideas on their heads.

His starting point for experiments on the heart was the ancient notion that systems in nature often operate in cycles. Seasons, day and night, the movement of the moon: all followed cyclical patterns. If that were so in the cosmos, it should also be so in the microcosm—the body. When he thought about blood he wondered if it, too, moved in a cycle.

So far he was operating as a classic philosopher, but then he carried out some practical experiments. By counting the number of times the heart beats, and making crude estimates of how much blood the heart could contain, he realized that it was far too much to be on a one-way journey to the rest of the body. By occluding the veins and arteries supplying and leaving the hearts of animals as diverse as deer and snakes, he saw that blood flowed into the heart from the veins and left it in the arteries.

In a leap of imagination he then proposed that the blood passed from the arteries back into the veins within the organs. It took another half century before Italian Marcello Malpighi (1628–1694) saw capillaries through a primitive microscope and confirmed that Harvey's ideas were correct.

Harvey's work acts as a bridge between ancient philosophy and the mode of thinking that became known as science. He realized that to further medicine, you needed to develop theories and then test them with experiments and measurements to see if they hold true.

Born
1578, Folkestone, England

Died
1657, London, England

Harvey was the first scientist to use experiments and data analysis to conclude that blood circulated around the body. It was a concept that led to a radical rethink of the nature of life.

Carl Linnaeus
Classifying Living Things

Born
1707, Råshult, Sweden

Died
1778, Uppsala, Sweden

As one of a group of eighteenth-century "natural theologians," Swedish-born Carl Linnaeus had an underlying belief that since God had created the world, everything in it should be ordered and capable of being understood. This desire for organization and a deeper understanding of things, coupled with his natural scientific curiosity, led to Linnaeus becoming the father of modern taxonomy.

At the time Linnaeus was working, the mechanism of sexual reproduction in plants had only just been discovered. Fascinated by this, he studied the shapes and functions of reproductive organs from the many plant specimens that he collected. This enabled him to create a way of classifying each plant that was based on the number and shape of each one's reproductive organs—specifically the male stamens, which produce pollen, and the female pistils, which collect the pollen and produce seeds.

Linnaeus's concentration on the sexual organs of plants produced some strange results, for example it grouped fungi, mosses, lichens, algae, and ferns together, because they had no obvious sexual organs. Consequently his approach was criticized and later supplanted by systems that took into account many other physical features of a plant specimen.

One major feature of Linnaeus's system that has survived is his binomial method that uses a two-part name for each species, for example *Homo sapiens*. The first part refers to the larger group, or "genus," of organisms, with the second part specifying an individual species within that group. This was similar to work first pioneered by Aristotle (384–322 BCE), but Linnaeus took the idea further and grouped the genera into "orders," subdivided these orders into "classes," and these into "kingdoms."

New genetic findings at the end of the twentieth century led biologists to switch some animals and plants between groups but although the tables of names have changed, it nonetheless confirms that Linnaeus's overall concept was a lasting one, which continues to be used in biological study.

The contribution that **Linnaeus** made to taxonomy goes far beyond contributing so-called scientific names to many of the world's plants and animals. He developed a system of classification whose grouping schemes now underlie many aspects of biology.

Claude Bernard
Pioneering Medical Research

Claude Bernard studied medicine in Paris, qualifying in 1839. Tutored by François Magendie, a leading physiology researcher who was also involved in animal research, Bernard later became Magendie's research assistant at the Collège de France. Bernard had great manual dexterity and an approach to designing experiments that was firmly based on scientific theory. He developed many new techniques and approaches in animal experimentation.

Bernard proposed what he called the "internal milieu" of the body. According to him, the body creates its own inner environment where fluids, cells, and organs work in equilibrium with one another in health. In illness this balance is disrupted.

One of Bernard's major discoveries, in 1848, showed how the pancreas secreted enzymes that break down fat, protein, and carbohydrates. Other experiments on the digestive system showed the presence of an enzyme in the gastric juices of the stomach and the alteration of carbohydrates into simple sugars before absorption in the gastrointestinal tract.

Bernard also carried out important work on toxicology, studying the poison "curare," used as a weapon by South American Indians. Curare acts where the nerve meets the muscle, stopping the nerve from contracting the muscle, thus causing widespread paralysis. Bernard used this, and similar studies on carbon monoxide and opium, to show that the action of poisons and drugs, are specific to targets within the body. Through animal experiments carried out in 1852 to 1853, he also showed the vasomotor effect of nerves, where they can either contract or dilate the blood vessels, so playing an important role in regulating the temperature of the body. Bernard's great work *Introduction to the Study of Experimental Medicine* was published in 1865.

In 1852, Magendie retired and Bernard took over most of his work at the Collège de France. In 1854, a professorship of general physiology was created for him at the Sorbonne. Bernard gained an Academy of Sciences award for his work on the nervous system, was elected a member of the Academie Française, and became President of the French Academy. When he died, of a kidney complaint, Bernard was given a public funeral, becoming the first-ever French scientist to receive this honor.

Born
1813, Saint-Julien-en-Beaujolais, France

Died
1878, Paris, France

More interested in medical research than practice, **Bernard** believed that physiology, pathology, and pharmacology were closely linked and that all three should be considered on par with physics and chemistry.

Karl Landsteiner

Understanding Blood Groups

Taking blood from one person and infusing it into a patient seems to be a relatively straightforward process, but when it was first tried recipients often had huge reactions against the donated blood. When Karl Landsteiner's scientific work unraveled the mechanisms that enable safe blood transfusion he also opened the door for the development of successful organ transplants.

Blood is composed of many different types of cell. Red blood cells carry oxygen from lungs to tissues and enable blood to transport carbon dioxide from tissue to lungs. White cells fight infections and platelets help blood to clot. These cells float in a liquid called plasma. If blood is allowed to clot, the plasma becomes serum—a liquid that has slightly different properties from plasma because the clotting factors have been used up.

In 1900, Landsteiner started to observe the way that blood responded when it was mixed. Mixing serum from one person with whole blood taken from another sometimes caused the red cells to clump together. In other combinations this did not occur. Landsteiner concluded that people came from one of three groups, which he called A, B, and C. A year later, one of Landsteiner's pupils discovered a fourth group—a set of people whose serum didn't clump the red blood cells of either A- or B-type individuals. He called this group O.

This clumping process occurs because blood plasma contains a series of large molecules called antibodies. These search out proteins that come from outside the person's body—usually as a defense mechanism against bacteria and viruses. However, these antibodies will also attack injected foreign blood cells—if they can spot that they come from outside the body. Blood groups exist because blood cells are covered with tiny proteins, and while people from group A all have the same version of that protein, those in group B have a different version. If group B blood is injected into a group-A person, the recipient's antibodies lock onto these newly arrived proteins and trigger a response that destroys the cell. People in group O produce blood that has none of these marker proteins; such people are therefore "universal donors"—that is, their blood can be donated to any person. This amazing discovery on Landsteiner's behalf not only made him the father of transfusion medicine but has also allowed us to go far beyond blood transfusion to complete organ transplant.

Born
1868, Vienna, Austria

Died
1943, New York, New York

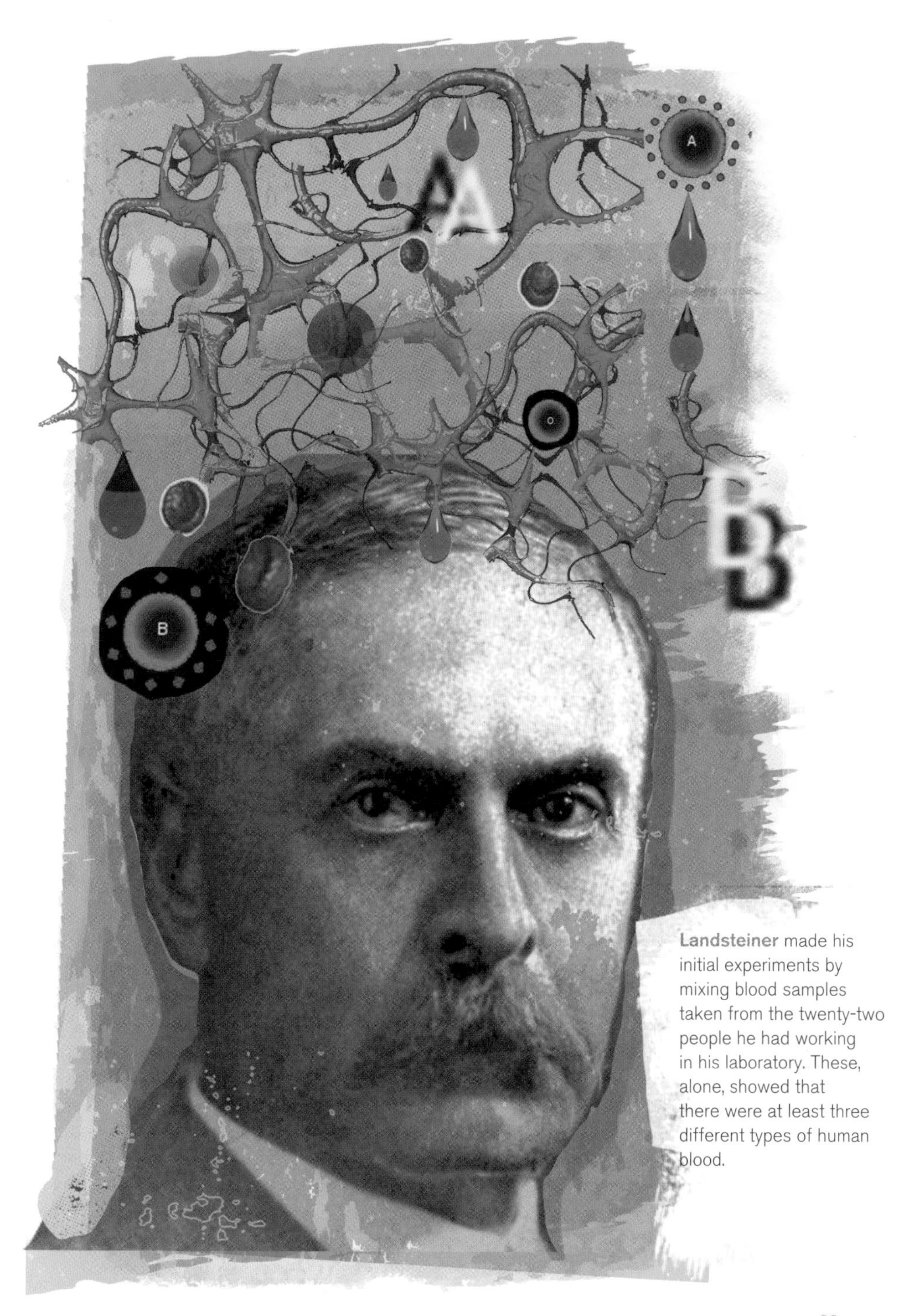

Landsteiner made his initial experiments by mixing blood samples taken from the twenty-two people he had working in his laboratory. These, alone, showed that there were at least three different types of human blood.

Carl Djerassi

Giving Birth to the Pill

As early as the 1920s, Austrian Professor of Physiology Ludwig Haberlandt (1885–1932) established that a chemical released by the ovaries could prevent pregnancy. The chemical—progesterone—was a hormone produced in large quantities by the ovaries of pregnant females. The problem was how to make use of this chemical, since it was difficult to purify from natural sources, hard to make artificially, and even then, was poorly absorbed into the body if taken as a tablet.

The solution came from an inedible species of South American yam. A researcher at Pennsylvania State College discovered that it produced vast quantities of diosgenin, a chemical that could be easily converted into progesterone. At this point Austrian chemist Carl Djerassi started synthesizing a range of different hormones from diosgenin. In 1951, working with a young Mexican chemist named Luis Miramontes, Djerassi synthesized 19-nor-17alpha-ethynyl-testosterone, called norethindrone for short. If taken in tablet form, this steroid hormone could be absorbed well and had all the power of progesterone.

Djerassi and his colleagues were excited about the product and were sure it would be useful in helping women with menstrual disorders, but its real impact became clear when a group of medical researchers in Puerto Rico and Haiti ran a clinical trial in 1956 involving 6,000 women. When this showed that norethindrone could control the pregnancy rate, the world woke up to the realization that a drug had been found that could empower women, giving them more control over their reproductive destiny.

Soon "the Pill" was born. Though it might be unrealistic to say that norethindrone led the sexual revolution, it would be equally unrealistic to say that the sexual revolution would have occurred at the same rate without it. The Pill brought about indisputable change in the economic and social status of women and has unquestionably improved women's health, by reducing pregnancies and miscarriages as well as the incidence of ovarian and endometrial cancers.

Born
1923, Vienna, Austria

Died
2015, San Francisco, California

When **Djerassi** and his colleagues first developed norethindrone, they didn't see it as a world-altering compound. Today, millions of women worldwide take the Pill every day.

1750

Edward Jenner *An Inquiry Into the Causes and Effects of the Variolæ Vaccinæ* (1798)

1800

1850

1855

Charles Darwin *On the Origin of Species* (1859)
Rudolf Virchow "Vorlesungen über Cellularpathologie in ihrer Begründung auf physiologischer und pathologischer Gewebelehre" (Cellular Pathology as Based upon Physiological and Pathological History) (1859)

1860

Gregor Mendel Verhandlungen des naturforschenden Vereins Brünn (Proceedings of the Natural History Society of Brünn) (1866)

1865

Robert Koch "Untersuchungen über Bakterien: V. Die Ätiologie der Milzbrand-Krankheit, begründet auf die Entwicklungsgeschichte des Bacillus anthracis" (Investigations into Bacteria: V. The etiology of anthrax, based on the ontogenesis of Bacillus anthracis) (1876)

1870

1880

Louis Pasteur "Traitement de la Rage" (Rabies) (1886)

1890

1920

Alexander Fleming On the antibacterial action of cultures of a penicillium, with special reference to their use in the isolation of B. influenzae (1929)

1930

Barbara McClintock "The Stability of Broken Ends of Chromosomes in Zea Mays" (1941)
Oswald Avery "Studies on the Chemical Nature of the Substance Inducing Transformation of Pneumococcal Types" (1944)

1940

1945

Crick and Watson "Molecular structure of Nucleic Acids: A Structure for Deoxyribose Nucleic Acid" (1953)

1950

1955

Genetics and Microbiology

The biological domain of the very small is a recent development in the history of science, but the discoveries made by the scientists studying microbiology and genetics have led to some of the greatest advances in human health as well as giving us a better understanding of life itself.

Evolution

Charles Darwin's celebrated publication of *On the Origin of Species* forms just one step in a chain of events marking humanity's struggle to make sense of the biological world.

In ancient Greece, Aristotle (384–322 BCE) and his student Theophrastus (later known as the father of botany) realized that they could place plants and animals into groups by comparing their physical features. This accompanied a realization that, on the whole, plants and animals could only produce offspring if they bred with other highly similar individuals—in other words, other members of the same species.

For over two thousand years the reason for this seemed obvious. God, or some creative influence, had generated all the species found on Earth and had made biological boundaries that ensured that species didn't get mixed up.

By the time that Charles Darwin (1809–1882) stepped aboard HMS *Beagle* and headed west to explore South America and the Galapagos Islands in the Pacific Ocean, people were already beginning to question this static idea of biological existence. Charles Lyell (1779–1875) had begun to show that looking at the structure of the Earth indicated that it may have existed for millions of years. This alone sparked questions of whether anything had changed significantly over that time.

On November 24, 1859, when Darwin eventually published his book, science took a massive step into the unknown—but a step that now underlies much of current biological thinking. The full title of his book reads *On the Origin of Species by Means of Natural Selection, or the Preservation of Favoured Races in the Struggle for Life.* In it he shows how studying a wide range of plants and animals, as well as the environments they were found in, led him to conclude that animals that are best fitted to their environment are most likely to breed and pass on their characteristics to the next generation. Over enough

"We can allow satellites, planets, suns, universe, nay whole systems of universes, to be governed by laws, but the smallest insect, we wish to be created at once by special act."

–Charles Darwin

time, he suggested, this process of gradual persistent change could generate individuals that were so different from their ancestors that they could no longer breed with non-changed members of the species. In other words, they developed to such an extent that they constituted a new species.

At the time Darwin was working, no one knew about DNA or genes and so the mechanism that allowed this passage of information from one generation to the next was a mystery. Austrian monk Gregor Mendel, however, was in the process of working out that physical characteristics were passed on in discrete units of information—what would become known as "genes"—and was making distinct progress in determining some of the mathematical and statistical rules that describe the way this happens.

The arrival of genetic technology now reveals more details of the underlying mechanisms. The discovery that most living organisms share similar basic housekeeping genes points to a common ancestry, and analysis of the differences within these genes gives an indication of what species have evolved from each other. Darwin would be fascinated.

Edward Jenner
Discovering Vaccination

In the eighteenth century no one knew that bacteria or viruses existed, but they did know about smallpox. This disease gave influenza-like symptoms, followed by a rash all over an infected person's body. The rash developed into pus-filled blisters, and many people developed infections in the kidneys and lungs before they died.

The only known immunity was to catch smallpox and survive. In China, doctors deliberately gave people small doses of the disease by grinding smallpox scabs or fleas that had fed on cows with cowpox, and blowing some of the dust into their nostrils. The idea slowly spread west. In the early 1700s English aristocrat Lady Mary Wortley Montague came back from a trip to Turkey having seen women in the Ottoman court making small graze marks on children's arms and wiping the area with smallpox scabs. Impressed with the idea, she had her young children treated in this way. Although the technique had successes, it wasn't perfect. One in fifty people died as a result of such treatment, and it occasionally triggered smallpox outbreaks.

It was generally believed that milkmaids were protected from the disease, because those who had caught cowpox did not succumb to smallpox, and on May 14, 1796, Edward Jenner performed a vital, if terribly risky, experiment. He made two half-inch scratches on the arm of an eight-year-old boy and wiped a cowpox scab over the wound. The scab had originally come from the hands of a local milkmaid, Sarah Nelmes. Six weeks later, Jenner exposed the boy to smallpox. He did not become ill.

The process worked for both the Chinese and for Jenner because the viruses causing cowpox and smallpox are remarkably similar. When exposed to cowpox viruses the person's immune system developed molecules that are ready and waiting to fight the virus. These same molecules, however, were equally capable of fighting smallpox viruses, so the person developed resistance to both diseases. Vaccination had arrived.

By 1800 some 100,000 people throughout the world had been vaccinated against smallpox, and in the twentieth century the World Health Organization made a concerted effort to wipe out the disease. On October 27, 1977, Ali Maow Maalin, a twenty-three-year-old hospital cook in a small Somali village called Merka, became the last person to catch smallpox in the wild. For the rest of the world, smallpox had been consigned to history.

Born
1749, Berkeley, England

Died
1823, Berkeley, England

The father of immunology, **Jenner** was a pioneer of vaccination. The word itself comes from his experiment with the cowpox virus, and is taken from the Latin *vacca*, meaning "cow."

Charles Darwin
Developing Evolutionary Theory

When Charles Darwin concluded that new species had evolved from older ones by a process of natural selection, he created the theory of evolution and turned the world of biology, and theology, on its head.

In 1831, Darwin set sail on HMS *Beagle* as the companion of the ship's twenty-six-year-old captain Robert FitzRoy (1805–1865). On the voyage they visited the Cape Verde Islands, the South American coast, the Strait of Magellan, the Galapagos Islands, Tahiti, New Zealand, Australia, the Maldives, and Mauritius, before returning to England in 1833.

While traveling, Darwin read Charles Lyell's (1797–1875) *Principles of Geology*, which argued that the world was being shaped and reshaped by ongoing geological forces. This ran against the accepted view that the world had been created a long time ago, and had only changed occasionally as a result of major natural events such as catastrophic floods.

Arriving in South America, Darwin found fossil evidence that seemed to indicate some form of progression from simple to complex life forms. In addition, on February 20, 1835, Darwin experienced an earthquake on the southwest coast of South America, which lifted the land between three and ten feet. Later he found fossilized seashells high up in the mountains and wondered whether numerous previous quakes had driven them there. If so, this would support Lyell's theory that the earth was changing constantly.

From September to October 1835 Darwin visited the Galapagos Islands, which as far as he could see had been recently created by volcanic action. He was therefore surprised by the diversity of life that he found on the islands and, in particular, by the variety of species of finches, each with differently shaped beaks.

Darwin concluded that different species developed through a gradual process of evolution, and presented his theory to the Linnean Society of London on July 1, 1858. Just over a year later he published *On the Origin of Species*, perhaps the most influential book in the history of science.

Born
1809, Shrewsbury, England

Died
1882, Downe, England

On his groundbreaking trip to the Galapagos Islands, **Darwin** observed how the beaks of different species of finches were highly specialized depending on the food available on each island—a result, he argued, of natural selection.

Rudolf Virchow
Founding Cellular Pathology

Born
1821, Schivelbein, Pomerania (now Germany)

Died
1902, Berlin, Germany

Suspicious of the "germ theory" put forward by his contemporary, Louis Pasteur, Rudolf Virchow believed that disease came from within the body, and not from external infectious agents. In 1859, he published *Cellular Pathology as Based upon Physiological and Pathological Histology*, in which he claimed that the cell was the basic unit of life, capable of reproducing itself. A radical view for its time, it went against the popular view that life somehow arose spontaneously.

Virchow studied medicine in Berlin and gained a junior post at the city's leading hospital, the Charité. He subsequently moved to Würzburg, where he became Professor of Pathological Anatomy. In 1856, he returned to the University of Berlin to take up a post created for him at the newly established pathology institute.

Using developments in microscopy—such as the use of stains for tissues and cells, and the microtome, which cuts very thin slices of tissue for study—Virchow promoted the use of the microscope to advance the study of pathology. He described leukemia, a group of blood cancers, in 1845, and was also one of the first to study inflammation, embolism, and thrombosis (the last two being abnormal formations of blood clots) at the cellular level. He was the first to put forward the idea that a venous thrombosis (a blood clot [thrombus] that forms within a vein) in the leg could break off and travel to the lung, forming a potentially fatal embolism.

Virchow was greatly influenced by the work of Mathias Schleiden (1804–1881) and Theodore Schwann (1810–1882), German biologists working on the importance of the cell in biology and medicine. Schleiden claimed that plants were made up of cells, while Schwann had discovered that cells were the basic building blocks of all the animal tissues he studied. Their research led Virchow to believe that disease began in the cells.

He argued that disease and its symptoms either resulted from an abnormality within the cells, or from the cells' response to some kind of stimulus. His work revealed the cellular basis of cancer, which is still the basis of modern views of the disease, where one abnormal cell multiplies to form a tumor.

As well as his work in the field of pathology, Virchow was influential in improving public health in Germany: He believed that a doctor could be a vehicle for social reform, and as a member of Berlin City Council, he advised on a number of issues, including sewage disposal, school hygiene, and the inspection of meat.

Virchow's work on the cellular basis of disease showed that while damaged cells are normally eliminated by the body in a process called apoptosis, cancerous cells can evade this system and multiply freely.

Gregor Mendel
Understanding Genetics

Although their lifespans overlapped, Gregor Mendel and Charles Darwin (1809–1882) never met. But while Darwin was developing a theory of evolution, Mendel found statistical proof that plants and animals pass physical characteristics from one generation to the next.

From the simple observation of livestock and crops, people knew that the offspring of plants or animals showed mixtures of their parents' physical characteristics. In his first paper, *Experiments in Plant Hybridization*, Mendel wrote that he wanted to discover a "generally applicable law of the formation and development of hybrids."

He described how he had bred a type of pea, *Pisum sativum*. The plant came in many strains that when bred in isolation produced identical offspring, but gave more complex results when mixed to produce hybrids. Crossing a plant that produced smooth peas with another that produced wrinkled peas, gave offspring that all had smooth peas. When these second-generation offspring were interbred, some of the new plants produced smooth, and some wrinkled peas. Intriguingly the ratio was almost perfectly three plants with smooth peas for every one with wrinkled peas. He then looked at the way different colors were passed from one generation to the next and found that the same rule applied, and noted that inheritance of the color of both the plant's flowers and its peas occurred independently of pea shape.

But it took genius to make sense of these results. Mendel realized that such phenomena could only occur if three things were happening in the process of reproduction: first, characteristics must be carried from one generation to another by means of some physical "element"—we now call this element a "gene." Secondly, each characteristic must be recorded in the cells on a pair of these elements. Thirdly, Mendel concluded that some of the characteristics were dominant over others.

Mendel was certain that he had made an important discovery and was disappointed that others didn't share this view. Recognition only came some years after his death, in 1900, when three other scientists, Dutchman Hugo de Vries (1848–1935), German Carl Correns (1864–1933), and Austrian Erik Tschermak (1871–1962), read Mendel's work. They had each unknowingly repeated his work and discovered similar rules—but Mendel had gotten there first.

Born
1822, Heinzendorf, Austria

Died
1884, Brünn, Austrian Empire [now Brno, Czech Republic]

Mendel cultivated 28,000 pea plants in his bid to find a law for the formation and development of hybrids. In doing so he established that each physical characteristic of the plant was inherited separately: an early breakthrough in genetics.

Louis Pasteur
Discovering Germs

Working on the chemistry and optical properties of asymmetric molecules, Pasteur came to the conclusion that although many molecules exist in two forms that are mirror images of each other, if they have been produced by a biological process only one of the forms will be present. If both forms are present, a physical process most probably generated the molecules. Asymmetry differentiates the organic world from the mineral world. In other words, asymmetric molecules are always the product of life forces.

This discovery had an unexpected use when he was asked to solve a problem in a factory that was fermenting beet to produce alcohol, because on some occasions the factory instead generated lactic acid. At the time, fermentation was seen as a chemical process that occurred when you brought the right ingredients together: Sugar broke down into alcohol as a result of some inherent destabilizing vibrations. The yeast cells found in wine were thought to play no role in the process.

Pasteur realized that the lactic acid crystals he observed were all of one type and correctly concluded that they had been created by living organisms. He then realized that while healthy, round yeast cells generated alcohol, lactic acid was being produced by small, rod-like micro-organisms that we now know were bacteria. The solution to the factory owner's problem was to keep everything clean, but the repercussions were that Pasteur realized that microscopic "germs" could have a significant impact on life.

Later, Pasteur was called in to solve problems in the French silk-worm industry. The worms were either dying, or failing to spin silk. He found that healthy worms became infected when they nested on leaves used by infected worms. Without fully understanding what was going on, he solved the problem by recommending certain conditions of temperature, humidity, ventilation, and quality of food, as well as husbandry techniques that kept newly bred worms away from older ones.

Together this would probably have placed Pasteur in the history books, but it was when he realized that many common diseases, including cholera, diphtheria, scarlet fever, syphilis, and smallpox were caused by microbiological agents—or germs—that his place in history was assured.

Born
1822, Dole, France

Died
1895, Saint-Cloud, France

Pasteur showed that injecting heat-damaged bacteria into people made them immune to the disease caused by healthy bacteria. By doing this he invented modern vaccination and triggered a massive advance in the fight against disease.

Robert Koch
Linking Bacteria with Disease

Born the son of a miner, Robert Koch always loved biology, and while studying anthrax he became the first person to demonstrate that specific bacteria caused specific diseases.

Koch was influenced by Professor of Anatomy Jacob Henle (1809–1885), who believed that living, parasitic organisms caused infectious diseases. Koch set up a small laboratory, where he started studying anthrax, a disease that was rife among the farm animals in the area. He had a hunch that it was associated with a type of bacterium that had recently been discovered.

He took slivers of wood and spiked them into the spleens of animals that had died of anthrax, and then spiked them into mice. The mice became infected and died. Mice spiked with wood covered with a healthy animal's blood were unaffected. Clearly something in the infected blood was transmitting the disease.

But was it bacteria or something else that killed the mice? To answer this, Koch developed ways of culturing bacteria and ensured that he could select some that were several generations away from any that had come from an infected animal. These still caused anthrax. The results showed for the first time that it was these bacteria, and nothing else, that caused the disease.

Recognition among the scientific community led to Koch being given laboratory space in Berlin. Here he showed that you could grow bacteria on solid surfaces such as potato and on agar kept in a Petri dish. As Koch studied, he came up with a list of four features that a microorganism must satisfy before it can be definitely linked to a disease. "Koch's postulates" state that you must be able to:

1. *Isolate the organism from every animal that has symptoms of the disease*
2. *Propagate the bacteria in a laboratory*
3. *Reproduce the disease by injecting the organism into a suitable recipient*
4. *Re-isolate the organism from this recipient*

As his work developed, Koch traveled widely, investigating diseases in various parts of Europe and Africa. When he died, infectious disease had lost some of its mystery and was now the target of scientific research.

Born
1843, Clausthal, Germany

Died
1910, Baden-Baden, Germany

Koch carried out controlled experiments using Petri dishes to prove that there could be no doubt that bacteria were the root cause of certain diseases.

Oswald Avery
Identifying the Transforming Principle

Oswald Avery was among the first molecular biologists to suggest that DNA played a significant role in the "transforming principle," which allowed dead bacteria to pass its virulent properties on to live bacteria. His research stimulated much interest in DNA and led to the discovery that DNA carries the "life blueprint" for all living organisms.

Avery's real interest was research in microbiology, with a focus on the microbes responsible for tuberculosis and pneumonia. He was intrigued by findings reported in 1928 by Fred Griffith, a microbiologist working in London. Griffith was working with pneumococci, one of the organisms that could cause pneumonia. There were two types of pneumococci, which Griffith called "rough" and "smooth" after the appearance of their colonies when grown on agar medium. While the rough bacteria were innocuous, the smooth ones were virulent; a carbohydrate "coat" around each microbe allowed them to evade the immune system and therefore cause disease. When Griffith mixed live rough pneumococci with dead smooth pneumococci and injected the combination into lab mice, he found it was lethal. Something, which he called the "transforming principle," had passed from the dead bacteria, conferring its virulent properties on the live bacteria.

Avery spent years trying to pull the transforming principle out of his experimental mixtures in order to determine its chemical identity. He used enzymes to chop up various cell components in order to find the transforming principle. If an enzyme removed a cell component and the transforming property was retained in the experiment, then that component could be eliminated. By the 1940s, it was known that chromosomes carried genetic information, a "blueprint" of the biochemical and physical characteristics of an organism. Chromosomes are composed of roughly half protein and half deoxyribonucleic acid (DNA). At the time, protein was considered the more significant part by biochemists, as it is more complex than DNA.

But in 1944, Avery and his colleagues, Colin McLeod and Maclyn Macarty, announced that DNA was the famous transforming principle. Scientists later found that DNA is the basis of all genes, carrying the life blueprint in all organisms from bacteria to humans.

Born
1877, Halifax, Canada

Died
1955, Nashville, Tennessee

Avery discovered that it was DNA in chromosomes that was the carrier of genetic information, a breakthrough that paved the way for modern genetics.

Alexander Fleming
Discovering Penicillin

When Alexander Fleming was born, disease-causing bacteria had terrifying power simply because medical treatments were unreliable at best. At that time, a simple scratch from a rosebush could prove enough to kill someone. Fleming therefore started looking for substances that could kill bacteria without harming animal tissues.

Fleming had trained to be a doctor at St. Mary's Medical School, London, and was soon working with vaccine pioneer Almroth Wright (1861–1947). Hopes of finding chemical cures for diseases had been raised in 1909 when Paul Ehrlich (1854–1915) found a chemical that could treat syphilis. Having tried hundreds of compounds, "salvarsan," the six hundred and sixth, worked. Fleming soon became one of the very few physicians to administer salvarsan in London.

Fleming later discovered that tears contain lysozyme, a biological molecule that breaks chemical bonds within the cell walls of some bacteria, causing them to burst. It was a significant discovery, but didn't work on all bacteria. Then, in 1928, while working on the influenza virus, he made a chance observation. Glancing over a set of discarded Petri dishes, he noted that an area around a growth of mold was cleared of bacteria. He wondered if the mold could be producing a chemical that killed the bacteria, so took a sample to test. He discovered that this mold was a member of the *Penicillium* family and that it did indeed release chemicals that killed bacteria.

These chemicals turned out to inhibit one particular step in the biochemical process used by many bacteria to build their cell walls. Without this process, the bacteria burst as they tried to grow. Penicillin is called a bacteriocidal agent because it actively kills growing bacteria, though it must be noted that it doesn't work on all bacteria, as others use a different biochemical process for building their cell walls.

Although Fleming discovered this chemical, it was not until later, during World War II, that New Zealander Howard Walter Florey (1898–1968) and an interdisciplinary team that included chemist Ernst Boris Chain (1906–1976) succeeded in extracting penicillin and showing that it had great promise as a treatment.

Born
1881, Lochfield, Scotland

Died
1955, London, England

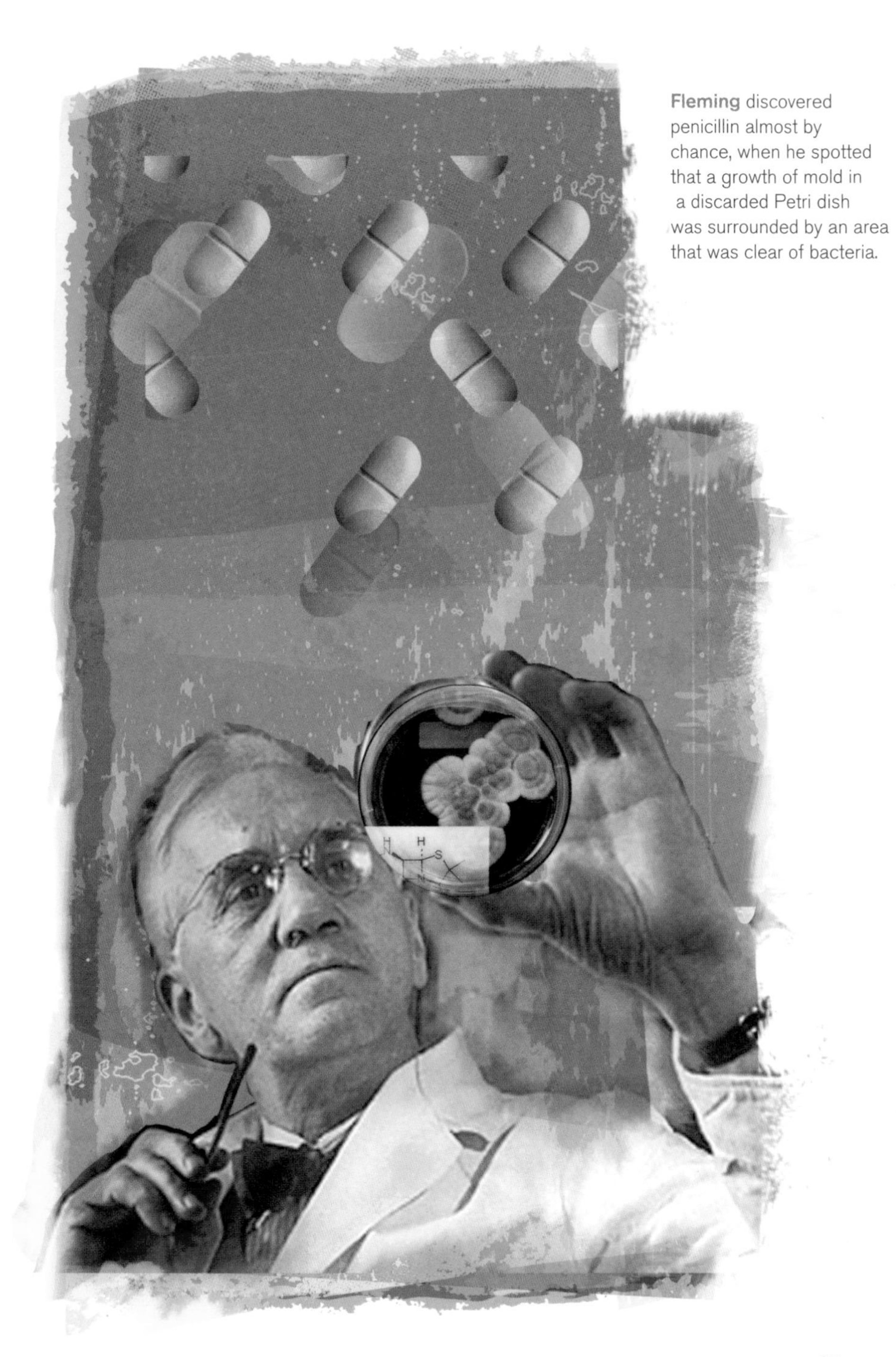

Fleming discovered penicillin almost by chance, when he spotted that a growth of mold in a discarded Petri dish was surrounded by an area that was clear of bacteria.

Barbara McClintock
Discovering the Movement of Genes

Born
1902, Hartford, Connecticut

Died
1992, Huntington, New York

Some early nineteenth-century scientists believed that traits such as the color of a flower and the shape of a pea could be inherited separately, because, during the cell division that generates pollen and seeds, parts of the various chromosomes broke off and swapped with each other. At the time there was no evidence to support this.

In 1927, Barbara McClintock began work with Harriet Creighton (1909–2004) at Cornell University in New York, to identify the ten different chromosomes present in maize cells. Then, working with a particular strain of maize, *Zea Mays*, that had a mutated chromosome number nine, the two researchers showed that parts of the chromosome did indeed swap over during the process that generates reproductive cells.

Next, working with Lewis Stadler (1869–1954) at the University of Missouri, McClintock started studying maize that had been exposed to X-rays. Peering into their cells with her microscope, she identified ring chromosomes. These she correctly suggested were parts of chromosomes that had been broken by radiation and subsequently fused to form a ring. Chromosomes occur in cells in pairs, and McClintock saw that chromosomes could break and fuse with the other member of the pair, before being ripped apart when the cell divided—a phenomenon that became known as the breakage-fusion-bridge cycle.

Moving to Cold Spring Harbor, McClintock discovered a bizarre genetic behavior in some of her breakage-fusion-bridge strains. Certain genes moved from cell to cell during development of the corn kernel. When she presented this data at a meeting in 1951, she was greeted with silence and some derision. As she said in her acceptance speech for the 1983 Nobel Prize for Physiology or Medicine, "Because I became involved in the subject of genetics only twenty-one years after the rediscovery, in 1900, of Mendel's principles of heredity . . . acceptance of these principles was not general among biologists."

Many scientists still believed that the function of a gene depended on where it was in a chromosome. If that were the case, then genes just couldn't jump around; or if they did, they would cease to function. McClintock's data only made sense if genes could move around on their own but still function. In time McClintock was proved right, and her theory of the nature of genes underpins current genetics.

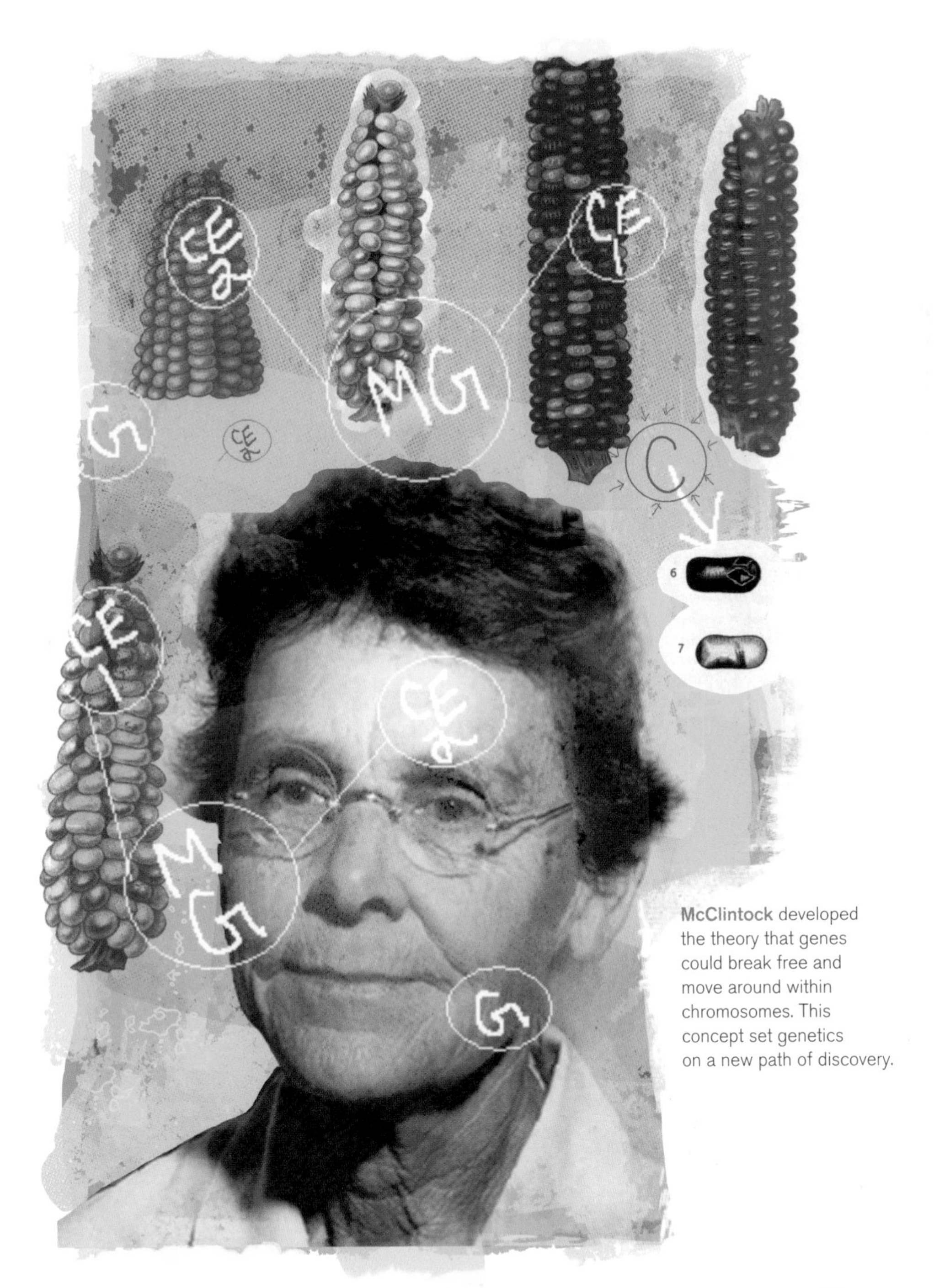

McClintock developed the theory that genes could break free and move around within chromosomes. This concept set genetics on a new path of discovery.

Crick and Watson
Uncovering the Structure of DNA

When they first met in Cambridge, England, in 1951, Francis Crick and James Watson shared a common curiosity: how to fathom the structure of a biological molecule called deoxyribonucleic acid (DNA).

By this time, scientists were sure that the information about the structure of an organism was stored in a cell's nucleus, passed on from cell to cell as an organism grows, and from parent to offspring in eggs and sperm. There was increasing evidence that the material inside the cell enabling this to occur was DNA, but a fundamental mystery remained: how could any biological molecule store enough information to guide the development of cells, organs, and indeed entire organisms, in a cell that is invisible to the naked eye?

Crick and Watson started to consider the structure of DNA. Their interest was enhanced after Watson attended a seminar in London where researcher Rosalind Franklin presented some cross-shaped X-ray diffraction pictures of DNA that indicated that the molecule had a helical structure.

From chemical analysis of DNA, Watson and Crick knew that it consisted of four different components called "bases," and made scale models of each base. They also realized that these bases could form chains, and so they tried building a triple helical arrangement of the components with the spiraling spine formed in the middle, and the rest of the bases pointing out.

The model didn't work, but in 1953 Watson got a sneak-preview of another of Franklin's X-ray diffraction images and realized that the pattern could best be explained by DNA being a *double* helix, with the two strands running in opposing directions.

Crick and Watson jumped to a radical conclusion: the two chains were effectively mirror images of each other. Each chain held equivalent information, and you could produce a second copy by separating the chains and using each as a template.

Genetics had come of age. The following years have seen a steady increase in the number of techniques that are enabling scientists to make sense of the genetic processes going on inside every living cell.

Francis Crick
Born
1916, Northampton, England
Died
2004, San Diego, California

James Watson
Born
1928, Chicago, Illinois

Crick and **Watson** discovered that the DNA molecule looks like a spiral ladder where the rungs are formed by base molecules, which occur in pairs. Each sequences of base pairs represents the genetic information stored in the DNA.

Glossary

Agar: Gelatinous material, derived from certain marine algae, that is used as a base on which to grow bacteria.

Atomic weight: The relative mass of an atom of an element compared to the mass of carbon-12.

Alchemy: The ancient predecessor of modern chemistry. The most well known goal of alchemy was the transmutation of any metal into either gold or silver. Alchemists also tried to create a panacea, a cure for all diseases and a way to prolong life indefinitely. A third goal of many alchemists was the creation of human life.

Algebra: Part of mathematics in which signs and letters are used to represent numbers.

Amino acids: The building blocks of proteins; the main material of the body's cells. Insulin is made of 51 amino acids joined together.

Anthrax: Bacterial infection that can cause a severe disease if it enters via the skin, but is often fatal if the bacteria or its spores are inhaled.

Big Bang: A theory of cosmology in which the expansion of the universe is presumed to have begun with a primeval explosion.

Black hole: An object with such high gravity that not even light can escape. Possibly formed when the most massive of stars die, and their cores collapse into a superdense mass.

Cancer: General term for diseases caused by abnormal and excessive growth and division of cells in the body.

Chyle: A milky fluid containing fat droplets that drains from the lacteals of the small intestine into the lymphatic system during digestion.

Chromosome: A threadlike package of genes in the nucleus of a cell, made of DNA wrapped around supporting proteins, visible under a microscope.

Einstein's Cosmological Constant: To get a static universe, Einstein added an artificial term, his cosmological constant, to his field equations that stabilized the universe against expansion or contraction.

Electron: A negative charged particle that orbits the nucleus of an atom.

Electromagnetic radiation: A propagating wave in space with electric and magnetic components. Electromagnetic radiation is also used as a synonym for electromagnetic waves in general including, for example, light traveling through an optical fiber. Electromagnetic (EM) radiation carries energy and momentum, which may be imparted when it interacts with matter.

Enzyme: A protein or protein-based molecule that speeds up chemical reactions in living things.

Gene: The segment of DNA on a chromosome that contains the information necessary to make a particular protein.

Glucose: A simple sugar used as the body's basic fuel and derived from the carbohydrates in the diet.

Hormone: A chemical messenger within the body that is secreted by one type of cell and acts on another type of cell.

Molecule: A collection of two or more atoms held together by chemical bonds.

Prion: An infectious particle that does not contain DNA or RNA, but consists of only a hydrophobic protein; believed to be the smallest infectious particle known that transmits the disease from one cell to another and from one animal to another.

Quantum theory: The theory that describes laws of physics that apply on very small scales. The essential feature is that energy, momentum, and angular momentum come in discrete amounts called quanta.

Red shift: The light coming from distant objects is slightly redder than predicted. This red shift is best explained as a lengthening of the wave length of light caused by the universe expanding.

Resistor: A component of an electrical device that offers resistance to electrical current flow.

Scarlet fever: A disease that results from infection with a strain of Streptococcus bacteria, spread by respiratory droplets. It carries an erythrogenic (rash-inducing) toxin that causes the skin to shed itself

Serum: The clear, thin, and sticky fluid portion of the blood that remains after blood clots. Serum contains no blood cells, platelets, or fibrinogen.

Thalamus: Part of the lower brain that relays nerve impulses carrying sensory information from the spinal cord to the cerebral cortex.

Theory of special relativity: Einstein's theory that challenged and overturned Newtonian physics by showing how the speed of light was fixed, and was not relative to the movement of the observer.

Virus: Ultra-microscopic infectious agent that replicates itself only within cells of living hosts.

Yeast: Single-celled organisms that occur naturally, and are used in baking and brewing industries.

Index